NO ROOM FOR ERROR

Pete van der Spek

SPEARHEAD

Published by Spearhead
An imprint of New Africa Books (Pty) Ltd.
99 Garfield Road
Claremont
7700

(021) 674 4136
info@newafricabooks.co.za

First edition 2003

ISBN 0-86486-555-4

Editing by Sue Ollerhead
Cover design by Toby Newsome
Design and typesetting by Peter Stuckey
Printed and bound by ABC Press, Cape Town

Contents

Acknowledgements   iv

Preface   v

Chapter 1 Personal experiences   1

Chapter 2 Air shows   5

Chapter 3 Crop spraying and firebombing   19

Chapter 4 Mountains and weather   24

Chapter 5 Aerobatics   56

Chapter 6 Mid-air collisions   63

Chapter 7 Helicopters   67

Chapter 8 Fuel mismanagement   78

Chapter 9 Less common factors   84

Chapter 10 The CAA   112

Chapter 11 The experts   116

Dedication   119

Glossary   121

Acknowledgements

THE INFORMATION FOR this book comes mostly from the Civil Aviation Authority (CAA) Investigation Department in Pretoria, who kindly granted me permission to reproduce photographs from their case files. Without the kind and willing assistance of Klaus Schwerdtfeger, this book would not have been possible.

To Roy Downes and Willy Roets, thank you for your initial help. You were there when I first set out to write this book and urged me on – thank you.

To Sue, my wife, thanks for all your support and believing in me.

To guys like Karl Jensen, Scully Levin and Stu Davidson, who love flying and are the 'industry standard' – keep up the good work that you do to create a huge interest in aviation.

To the many aviation enthusiasts out there who said this book was a good idea – I hope you enjoy it.

Pete van der Spek
puds@telkomsa.net

Preface

I HAVE BEEN an aircraft enthusiast for as long as I can remember. However, it was the crash of the Boeing SAA *Pretoria* on take-off from Windhoek International Airport in 1968 that sparked my fascination for aircraft accidents.

I started this book about six years ago, slowly adding to it as I went along. I have been encouraged by many people along the way, many not even in the aviation industry, and I'd like to thank them all for their support.

This book is not about blaming anybody. It is a book that draws attention to the fallibility of human beings. It is about documenting the mistakes made by pilots – in some cases, at the cost of people's lives.

The main object of this book is to point out to the many pilots out there, both in the aviation industry and the private sector, the possible consequences of their actions if risks are taken. If my book can save just one life, then I will have achieved what I set out to do.

Reading through a small selection of the files at the Civil Aviation Authority (CAA), it struck me how often pilots do take chances. Of course, there are the genuine errors that do occur, but overall, the boldness and the attitude of some pilots is astounding.

The old adage, 'There are bold pilots, there are old pilots, but there are no old, bold pilots' is very true. In support of this, are wise words from experienced SAA pilot, Captain Darryl Lee: 'The only way to fly is to stay ahead of the aircraft.' In other words, do everything by the book (the book was written by the experts) and always anticipate your aircraft's next move and your next action.

I have selected a few examples of accidents caused by different factors, such as bad weather, inexperience, poor health and fuel mismanagement. There are also chapters on mid-air collisions, mountain mishaps, air show and helicopter accidents.

Every pilot thinks he is good. Whilst there is nothing wrong with that, many pilots become overconfident – and that's when accidents can happen.

So, here is my version of a selection of accidents obtained from the CAA files. In each case the views expressed are mainly mine as is the interpretation of what happened, although the details are from the accident files. Without the help of the CAA, this book would not have been possible.

The men and women of the CAA need a special mention. Much of their time is spent on detective work – combing through evidence for clues leading to the cause of a particular accident. At times it is an unpleasant occupation, but I think a most satisfying and valuable one. They solve most cases, sometimes under very trying conditions, as they piece together the final moments of an aircraft's journey. Their hard work goes a long way towards making our skies safer.

CHAPTER 1

Personal experiences

I HAVE ALWAYS had a great love of flying in aeroplanes. As a young boy, my very first flight was in a Tiger Moth piloted by my Uncle Tom, who was a WW2 veteran. We did a circuit and then a loop. It was the most wonderful moment of my childhood. I don't think I stopped talking about it for weeks.

From that day I grabbed every opportunity to fly – and have done so in many different types of aircraft, ranging from the military Dakota and the State President's Viscount, to light aircraft such as Maulles and Kudus. I have also flown in Alouette, Bell Jet Rangers, Puma and Augusta 109 helicopters, and of course all the commercial jets.

Nowadays, every time I fly in a commercial aircraft, I nag to go up to the 'pointy end' (cockpit), for the best view in the house. It helps to calm my nerves – a hangover from an accident in 1994 in which the pilot and I could have been badly injured, or even killed, had it not been for his excellent training.

Ever since that accident, I have refused any invitation from a light aircraft pilot, unless of course it is someone I know and trust implicitly.

I am not saying that most pilots are reckless, far from it. Most are very good. South Africa has good training facilities and the CAA ensures a high standard of pilots. Unfortunately, there are still too many pilots out there who are willing to take shortcuts. The system cannot expose them all.

Going for a flip

Date:	3 February 1994
Conditions:	Fine and calm
Aircraft Type:	Kudu C4M
Pilot Age:	35
Flying Hours:	881
Location:	Virginia Airport, Durban, KwaZulu-Natal

Synopsis
An ex-South African Air Force (SAAF) Kudu was landing at Virginia Airport when the left wheel appeared to seize, causing the aircraft to veer left. To avoid a more serious collision, the pilot flipped the plane, which landed upside down next to the runway.

Flight details
My friend Pottie, an ex-SAAF pilot , phoned me to say he was going for a short test flight in his employer's Kudu to try out the global positioning system (GPS), which had just been repaired.

> A **GPS** is a space-based, global, all-weather, radio positioning navigation system. It determines position location via signals from a satellite constellation system. It can determine a user's latitude, longitude and altitude.

He was due to fly his employers up the East Coast the next day. I remember him inviting me in his inimitable style: 'Pete, would you like to come for a quick flip?' If only I had known then that we would flip – literally!

We took off from Virginia Airport, flew up to Ballito on the North Coast and returned to Virginia. It was a perfect flight – till we touched down.

The approach was textbook stuff. Any tail-dragging aircraft should be landed in a specific way. Touchdown should occur on all three wheels simultaneously. The pilot should fly the aircraft in at just above stalling speed and then let it settle on to the runway. It is a difficult technique to learn, but once mastered, quite easy. The touchdown was perfect, with the stall warning coming on just before the aircraft touched down. The aircraft ran straight for a few metres, then Pottie gently started braking.

It was then that things started to go wrong. The aircraft veered left and was heading for the dense bush, bordering the sea-facing side of the Virginia airport. The bush is perhaps 10 metres from the runway, so there is not too much margin for error. I remember seeing the bush coming towards us at an alarming rate. Our speed was probably only 90 to 100 kilometres per hour, but it felt a great deal faster.

This is where Pottie's air force training kicked in! He lifted the left wheel off the ground, gently applying the right brake to try to bring the Kudu into line. He then decided that the only way to avoid ending up in the bush was to 'stand on the brakes' and flip the plane – which is exactly what he did.

There was a loud bang as we tipped over. Pottie immediately switched everything off and we undid our seatbelts and fell to the floor (roof). We crawled from the aircraft and were met by the airport emergency teams. The nose and propellor of the Kudu had been damaged and the left wing was just in the bush. The aircraft's tail had taken the brunt of the impact and the engine and airframe were so badly damaged that the aircraft was written off. Fortunately neither of us were injured, although there may have been a slight dent in Pottie's pride!

The inquiry into the crash found that the left brake was more sensitive than the right, which caused the Kudu to pull left. My own impression was that the left wheel seized for a few seconds, and by the time it unlocked we were heading for the bush.

3

The accident happened chillingly quickly. Time of touchdown to time of impact could not have been more than 5 or 6 seconds.

One cannot put enough emphasis on emergency training. Every pilot should take extra note of emergency procedures, as they could one day save his life. He has only seconds to react.

This photo indicates just how close the Kudu came to crashing into the dense KwaZulu-Natal bush.

CHAPTER 2

Air
shows

NOTHING CAN BEAT the thrill of an air show: the heady roar of the engines and the thrilling stunts playing to enraptured audiences. Unfortunately, air shows have also set the stage for tragic accidents. Spectators equipped with modern technology have been able to capture these incidents on film. I happen to have witnessed several of these accidents first-hand, some more high-profile than others. In this chapter I focus on the numerous causes of these accidents.

The golden rule at air shows is never to wander over the crowd line. The crowd line demarcates the spectator area and, although set back from the runway, is usually near the taxiway so that spectators get a good view of the aircraft. Before each air show, organisers convene a safety meeting detailing procedures, event schedules, crowd line details and essential safety concerns.

An accident in the Ukraine city of Lviv involving a Sukhoi SU-27, highlights the necessity for adhering to clear crowd line details. In this accident, the Sukhoi fighter jet was flying close to the crowd line when the pilot lost control of the aircraft. The jet cartwheeled into the crowd, killing 81 people and injuring many spectators.

Adherence to strict procedures should help to prevent such appalling tragedies.

Hot and high

Date:	8 October 1977
Conditions:	Hot and calm
Aircraft Type:	Brittan Norman Trilander
Location:	Lanseria Airport, Gauteng

Synopsis
The Trilander pilot was attempting to show the aircraft's capabilities while ignoring the dangers posed by 'hot and high' conditions. The aircraft simply ran out of height and crashed.

Flight details
The air show at Lanseria on 8 October 1977 promised to be an exciting event. Unfortunately, it was to be marred by one major incident.

An awareness of hot and high conditions is critical on the Highveld. Lanseria Airport lies at 4 517 feet above sea level (ASL). Such a high altitude should act as a red light to any pilot.

A Brittan Norman Trilander is a 'short take-off and landing' (STOL) aircraft. It is ungainly looking with a third engine mounted on the tail.

Hot and high is a term used to describe a situation in which an aircraft has to take off from an airport located at a high altitude (above 2 500 metres ASL) and in hot weather (above 26-28 degrees). Under these conditions, the air is much thinner than at sea level and the combination of high altitude and high temperature create flying conditions which require the pilot to take extra precautions.

After take-off, the pilot did a very steep climb and as soon as he got to the top, he did a wing-over. At this stage, he was almost over the crowd line. At the top of the next climb, he did another wing-over and dived down over the taxiway.

> A **wing-over** can be compared to a vehicle handbrake turn. When the aircraft slows down at the top of a climb, the pilot pushes the rudder all the way in and does a turn coming back on the same line as the climb out.

As he climbed out my immediate thought was, 'If he does another wing-over now, there's going to be trouble.' I followed the aircraft down with my camera and the picture produced here was the result.

The Brittan Trilander as it hits the ground near to the crowd line.

The aircraft hit the taxiway, immediately losing its fixed wheels and left engine. The third tail-mounted engine canted, but was still attached to the fuselage. The aircraft bounced back into the air, climbed to about 300 feet and then floated across the airfield, where it eventually hit the ground nosefirst. It bounced a further 50 metres, where it finally came to a stop. The aircraft was completely destroyed but the crew members, who sustained back injuries, were miraculously able to clamber out of the fuselage, which had broken in two. Fortunately, the aircraft came to rest on the opposite side of the airfield, away from the hangars and spectators.

So, what went wrong?

Quite simply, the hot conditions combined with an altitude of 4 500 feet created an air density which could not sustain the final manoeuvre. This accident shows the need for pilots to take both weather conditions and altitude into account when showing off an aircraft's capabilities.

One manoeuvre too many

Date:	3 May 1996
Conditions:	Fair, partly cloudy with a 5-knot wind
Aircraft Type:	Bellanca 8KCAB
Pilot Age:	62
Flying Hours:	1 364
Location:	Margate, KwaZulu-Natal

Synopsis

While practising formation flying for an air show the following day, the pilot of a Bellanca performed a flawed wing-over manoeuvre. He crashed the plane, killing himself and a passenger.

Flight details

The formation was due to consist of a three-aircraft team, but the third aircraft pulled out days prior to the event. After much discussion, the two remaining pilots agreed to carry on with the fly-past incorporating two aircraft instead of the original three.

The plan was to practise the manoeuvres in the general flying area and then return to Margate Airport to demonstrate the sequence to the event's safety officer.

The pilot of the Bellanca agreed that the pilot of the second aircraft would lead the formation and that they would communicate with each other on a dedicated frequency.

After six attempts at a wing-over, it became obvious that the pilot of the Bellanca was not familiar with the manoeuvre. The lead pilot questioned his method and speed as the Bellanca was becoming semi-inverted and losing too much height during the recovery. The lead pilot suggested that they cut the practice short and return to Margate Airport to discuss the sequence and execution on the ground.

During their return flight, the lead pilot worked the radio for both aircraft. He asked for clearance for a low fly-past followed by a full stop landing. The request was approved with a minimum height requirement of 200 feet AGL.

They descended to 250 feet, over the threshold of runway 22. The lead pilot continued down over the runway at 300 feet AGL and the Bellanca pilot confirmed that he was behind him to his right. This was their last communication.

On landing, the lead pilot was informed by witnesses that the Bellanca's pilot had attempted what appeared to be a stall turn. The aircraft became inverted and the pilot could not recover in time.

A witness to the accident, also a pilot, was standing at the threshold of runway 22 and saw the two aircraft approaching at low level. The Bellanca on the right pulled up sharply. It seemed as if the pilot was trying to perform a low level loop. It then disappeared behind a hill on the right of runway 22. Seconds later, the witness heard a dull thud as the aircraft hit the ground. There was no explosion.

A **loop** is an aerobatic manoeuvre in which an aircraft flies in a complete vertical circle. An outside loop, begun at the top of the circle, is considerably more difficult to perform, because the pilot encounters negative G-forces throughout the manoeuvre.

Another witness, who was standing in front of the control tower, watched the two aircraft fly past in a southerly direction. At the end of the runway the Bellanca pulled up steeply and started what looked like a loop, followed by a wing-over. It then came down vertically. When the aircraft disappeared from the witness' view, the motor was still running at full throttle. He said it did not seem out of control as it was headed for a 200-metre valley and he expected the aircraft to fly out the other side.

In fact, the aircraft collided with the steep eastern slope of the valley. There were just 5 metres between the first impact marks and where the aircraft came to a halt – an extremely short distance at that speed.

Inspection by the CAA investigators found no evidence of a mechanical malfunction. The safety recommendations issued by the CAA relating to this case say it all:

'Although no rating from the CAA is necessary for aerobatics, all pilots who participate in or practise aerobatic manoeuvres, must plan and execute these manoeuvres as professionally and carefully as possible so as not to endanger the lives of other people and property.'

This is a classic case of a pilot thinking he had the correct manoeuvre for a particular sequence, yet getting it wrong. There is just one golden rule here – if in doubt, don't try it. Rather practise manoeuvres like this at altitude, where there is much more chance of recovery than at low levels.

The ground is very unforgiving!

The wreckage of the Bellanca lies in a valley close to Margate Airport.

Air show blues

Date:	10 October 1981
Conditions:	Good
Aircraft Type:	Pitts Special
Pilot Age:	49
Flying Hours:	15 400
Location:	Lanseria Airport, Gauteng

Synopsis

A pilot was putting a Pitts Special through its routine at an air show. When he came out of an inverted spin, he was over the crowd line with little airspeed and not enough height. He was forced to select a piece of open ground and crash the Pitts.

Flight details

On the day of the air show, I happened to arrive at the same time as the Pitts pilot. He didn't know me, but I recognised him as he was a well known aviation personality at the time. We struck up a conversation which ended with me wishing him a good flight!

Neither of us could have foreseen that before the day was out, he would be in intensive care fighting for his life.

The pilot began his routine ten minutes earlier than his scheduled slot, climbing to approximately 4 000 feet. He then went into an inverted spin which apparently was part of his routine. After approximately eight spins he attempted to pull out and appeared to be experiencing difficulty. At 500 feet above ground level he managed to pull out but was facing the wrong way, towards the hangars and the crowds. He banked and pulled up to get away from the crowd line.

Then the aircraft seemed to stall. He was right over the car park with nowhere to go. He put the nose of the aircraft down and aimed towards an open piece of land. He had no choice but to crash-land the Pitts right there.

The aircraft struck the ground slightly nose-down, banking to the right. It collided with a two-metre fence and came to a standstill 16 metres further on. Though the safety harness was properly engaged, the shoulder harnesses were completely torn out from the anchor points due to the severity of the crash. As a result, the pilot sustained severe back, face and leg injuries. The Pitts was written off.

I believe that the pilot's actions showed great presence of mind and courage. While he didn't have many choices, he averted a terrible accident by keeping his head and aiming directly for an open piece of land. The number of lives that could have been lost had he come down in the parking area doesn't bear thinking about.

The pilot deliberately crash-landed the Pitts next to a parking area, avoiding injuring or possibly killing spectators at an air show.

Air show pressure

Date:	8 August 1992
Conditions:	Good
Aircraft Type:	Extra EA 300
Pilot Age:	43
Flying Hours:	1298
Location:	Swartkops Air Force Base, Gauteng

Synopsis
At a combined air force/civilian air show, the pilot of an Extra EA 300 aerobatic aircraft crashed his aircraft during a manoeuvre. The pilot was killed.

Flight details

Several factors played a role in this horrific, fatal accident. The pilot was under extreme pressure to perform well and to finish his routine before the next aircraft appeared for its time slot.

The pilot was only given permission to participate in the show on 6 August, just two days before the event. He was unable to attend the safety briefing that day at 13h00, as he had to have the aircraft's faulty smoke system repaired. The pilot did however approach the chief air traffic control officer (CATCO) that day to enquire about his time slot and to get final details. His friend, who was to act as the ground co-ordinator, accompanied him. The CATCO informed him that his time slot was 09h19 to 09h27. He stressed that this had to be strictly adhered to as a Boeing 707 tanker and four Mirages were expected over the airfield immediately after his display. The pilot requested permission to operate on a dedicated VHF frequency of 122.6 MHz. This was to enable him to speak directly to his friend who was to direct his display.

At the last minute, the pilot was told that his slot would be delayed by four minutes. The pilot took off at 09h23 and immediately started his routine. A witness in the control tower remarked that the pilot was flying close to the crowd and was too low. According to the ground co-ordinator, the pilot flew at 15 feet above ground level and then pulled up and executed a series of vertical flick rolls, up to a height of approximately 300 feet.

> A **roll** is an aerobatic manoeuvre in which an aeroplane rotates completely around its longitudinal axis. A **vertical flick roll** is a variation of the roll.

The aircraft then half-rolled and dived, flying almost parallel to the runway until it impacted with the ground. As the aircraft slid over the runway it started to break up and burst into flames. Despite the fact that fire tenders and medical help were on the scene within 40 seconds, the pilot sustained serious burns to 60 per cent of his body. He was airlifted to a nearby hospital where he died from his injuries early the next morning.

The remains of the cockpit area of NCK. Notice how the cockpit is still largely intact, although badly damaged by fire.

An aerial view of NCK, showing how it skidded before it came to rest.

Below is a transcript of the communication between the pilot and the ground co-ordinator during his sequence:

09:26:02	**Pilot**	OK I'll do a tail slide.
:04	**Friend**	OK.
:05	**Pilot**	Keep talking, see if it grabs your fancy, hey?
:14	**Pilot**	OK baby.
:15	**Friend**	Can't hear you.
:17	**Pilot**	I'm talking to myself.
:34	**Pilot**	How does the negative ...
:50	**Pilot**	Tell me, am I too far away from the crowd?
:53	**Friend**	You keep disappearing Nick, come a bit closer.
:54	**Pilot**	OK.
:56	**Friend**	High as well, hey?
:59	**Pilot**	Say ...
09:27:01	**Friend**	You keep going very high as well.
:03	**Pilot**	OK.
:20	**Friend**	It's lovely.
:25	**Pilot**	Going to do a flick-flacker.
:28	**Friend**	OK.
:36	**Friend**	Too slow, hey?
:39	**Pilot**	I'll put it off here. [Smoke]
:41	**Friend**	OK.
:50	**Friend**	Very nice, do a knife-edge.
09:28:02	**Pilot**	My pair of glasses keeps coming off.
:05	**Friend**	That's lovely.
09:28:06	**Pilot**	Knife-edge coming up.
:08	**Friend**	Keep it low, hey?
:10	**Pilot**	OK.
:12	**Pilot**	I'll do it down the taxiway.

:25	**Friend**	Very nice.
:26	**Friend**	Do it again, come past with your face to the crowd all the time.
:33	**Pilot**	OK.
:34	**Friend**	Is that ...?
:35	**Pilot**	Yea, that's me.
:37	**Friend**	(Double transmission) – two minutes.
:39	**Pilot**	Two minutes – OK, knife-edge coming up.
:48	**Friend**	Very nice.
:50	**Pilot**	OK, what do you want now?
:52	**Friend**	Want you to put your smoke off, it's throwing flame, throwing flames out.
:56	**Pilot**	OK, it's off.
:58	**Friend**	[Expletive]! Do another of your low level, because they like it. I don't know where the Jumbo is, apparently it's on its way.
09:29:03	**Pilot**	OK.
:09	**Pilot**	OK, here comes low low.
:16	**Pilot**	OK.
:16	**Friend**	OK.
09:29:20	**Pilot**	Hell, I keep missing that (flick) point.
:23	**Friend**	I noticed that you've got one minute till the Jumbo. I think maybe do one more of that, cause as you can hear, the crowd is clapping for you and then call it off. Is the tower going to talk to us?
09:29:54		*The first radio calls to the crash tender were made.*

When one examines this accident, several points are worth considering:

- The pilot did not attend the pre-air show safety meeting and only had a briefing on the day.

- He did not seem to have any preset flight in mind but relied on his friend to direct him.

- He had the added pressure of a tight schedule before the next aircraft entered the airspace.

- He was flying too low for the manoeuvre he was attempting.

CHAPTER 3

Crop spraying and firebombing

CROP SPRAYING CAN be a dangerous business, due to the fact that it has to be done very close to the ground. There are numerous accidents in the industry each year as a result of pilots missing the obvious – fences, telephone wires, power lines and many other obstacles one would expect to find on a farm.

Accidents also happen simply because people do illogical, unpredictable things, like driving in front of an aircraft that is taking off.

Crop spraying pilots have special demands made of them. They must be able to fly continuously at very low altitudes – 10 feet above the crops at 150 kilometres per hour – sometimes even faster. They also require high levels of concentration.

Crop spraying aircraft are custom-built for manoeuvreability. They are not designed to do rolls or loops – but that doesn't stop some pilots from trying!

Crop spraying dangers

Date:	5 December 1996, at approximately 14h50
Conditions:	Good visibility and an 8-knot wind
Aircraft Type:	Ayres S2R-T34
Pilot Age:	42
Flying Hours:	6 313
Location:	Twyfelhoek farm, Ottoshoop District, North West Province

Synopsis
An Ayres S2R-T34 crop sprayer collided with a small truck while taking off from a farm airfield. All occupants of the truck were killed.

Flight details
This pilot had a lucky escape from injury when his crop sprayer struck a vehicle on the ground during take-off. The driver of the truck and four farm workers were not so fortunate. They were all killed in the accident.

The collision occurred at the intersection of the farm's runways 02 and 24. The aircraft was taking off from 02 loaded with 1 500 litres of chemicals. As it approached the intersection at an air speed of about 50 kilometres per hour, the truck suddenly appeared from the right on runway 24. At the last minute the driver saw the aircraft approaching, swerved to avoid a collision but was struck by the aircraft and came to rest about 28 metres further on.

A farm worker who witnessed the accident sketched the background to the incident. He had initially been instructed by the farm owner to transport the four workers to the site of operation. They were to act as 'human markers' for the crop spraying. The owner told the farm worker to avoid the runways as spraying was in progress and the aircraft was using the runway. He was told to follow the road around the airfield. The owner then decided to transport the workers himself and seemed to be in a hurry. Totally ignoring the instructions he had given to the farm worker, he took a shortcut across the runway, directly into the flight path of the crop sprayer which was taking off after reloading chemicals.

The land stretching from the threshold of runway 02 to the inter-section of runways 02 and 24 was overgrown with dense bush on the right-hand side, obscuring the truck driver's vision. By the time the truck driver caught sight of the plane, it was too late to avoid a collision.

The Ayres' propeller was ripped off by the force of the accident.

All that remained of the vehicle that was carrying the farm workers.

Firebombing!

Firebombing is similar to crop spraying in that it requires the same close-to-the-ground flying techniques and a similarly high level of concentration from the pilot.

Date:	3 August 1999
Conditions:	Fine with heavy smoke
Aircraft Type:	Ayres S2R-T34
Pilot Age:	31
Flying Hours:	In excess of 2 047 hours
Location:	Bhunya Area, Swaziland

Synopsis
While on a firebombing operation, an Ayres S2R-T34 lost a wing tip, causing the plane to crash near the fire site.

Flight details
Responding to an alert, an Ayres S2R-T34 crop sprayer undertook to dump water on a fire raging in a nearby pine plantation. The pilot's first pass was too high and he executed a go-around. On the second try, the massive load of water was dumped right onto the fire. On pulling up, the aircraft went into a spin manoeuvre, plummeted and crashed just beyond the fire.

Investigators from the CAA found the left-hand wing tip and an upper wing panel section strewn across the area near to where the load was dropped. Small paint flakes and debris were discovered on the flight path before the drop point, indicating that the wing tip had failed in flight.

This accident resulted in a detailed investigation, as there had been several markedly similar accidents involving this type of aircraft. The CAA pointed out that the Ayres S2R-T34 goes through a flightpath change just after dropping its load. After the load is released, the aircraft's centre of gravity drops, causing it to pitch upwards. This puts extra strain on the wings. Flying too fast can cause a severe reaction, as happened here. The pilot was forced to pull up as he was flying too quickly and would have entered instrument meteorological conditions (IMC) – thick smoke in this case, which would have prevented him from seeing the ground.

This probably caused the positive 'G' load to exceed the aircraft's limitations, resulting in the spin. Excessive turbulence caused by the raging fire could also have been a contributing factor.

Sadly, the pilot did not survive.

The obvious factor in this accident is speed. Although everything happens very quickly when crop spraying or firebombing, it is important to adhere to the manufacturer's stipulations and limitations. This is one time in aviation where speed kills.

> If a pilot enters conditions where there is little or no visibility, such as in cloud, heavy smoke or at night, he needs to use his instruments to navigate and fly safely. These conditions are referred to as **instrument meteorological conditions (IMC),** where the pilot cannot see the ground for a reference point. Flying by means of instruments is done according to **instrument flight rules (IFR).** Obtaining such a licence requires extensive exams and 200 hours of flying time.

The burnt wreckage of the Ayres S2R-T34 lies in a field in Swaziland.

CHAPTER 4

Mountains and weather

ANYONE WHO HAS hiked in mountains will know how quickly the clouds can come down and obscure the mountaintops. These conditions are extremely dangerous for pilots who are not instrument-rated, as reduced visibility means aircraft can easily become ensnared on towering peaks or rocky outcrops.

South Africa has had more than its fair share of accidents involving aircraft that have drifted off course in bad weather. One such accident which received extensive news coverage involved former South African cricket captain Hansie Cronjé, who died in a plane crash outside George on 1 June 2002. The Hawker Siddeley HS-748 carrying Cronjé and two others crashed into the rugged Outeniqua mountains in overcast, rainy conditions. News reports at the time claimed that the aircraft was unable to land in George due to poor weather conditions, and that the pilot tried to circle before landing when it crashed.

The wreckage of the Hawker was salvaged by a local businessman, who has displayed it in a museum in George. This high-profile case is being investigated by the CAA, and at the time of going to press, the results have not yet been made public.

Reading through the CAA files, I was amazed at the number of pilots who tried to navigate these conditions using visual flight rules (VFR).

A pilot who qualifies with a **private pilot licence (PPL)** may only fly in clear weather. A commercial pilot requires an **instrument rating** that allows him to use his instruments to navigate and fly safely at night or in bad weather conditions.

One such file involved a young, experienced ex-SAAF fighter pilot. This particular accident fired the nation's imagination when, on a trip from Richards Bay to Grand Central Airport in Midrand, his plane went missing and wasn't seen again for six months. The wreckage was only discovered when a climber tackling an unusually remote peak in the Drakensberg chanced upon it.

Many of the flights in this category are called controlled flight into terrain (CFIT). CFIT occurs when a pilot mistakenly believes he is flying with sufficient clearance from the ground. Thus, the aircraft is assumed to be under control by the pilot when it impacts with terrain, be it the ground or a mountain.

Why do these accidents happen? What causes a pilot to misread his instruments or in some cases, take the chance of flying visually when there is absolutely no chance of getting through safely? Perhaps overconfidence plays a big role, or maybe the biggest factor is a reluctance to say 'I have to divert'.

High workload

Date:	2 October, 1986, 05h23
Conditions:	Extreme cloud/mist
Aircraft Type:	Dassault Falcon 10
Pilot Age:	43
Flying Hours:	6 875
Location:	Haenertsburg, Limpopo Province

Synopsis
On a flight from Lanseria near Johannesburg to Tzaneen in Limpopo Province, the aircraft lost transmission on descent. It crashed into a pine plantation in Haenertsburg, killing the pilot, his co-pilot and two passengers.

Flight details
The pilot requested direct clearance from Lanseria to Tzaneen, flying at 33 000 feet. The flight time given was 30 minutes. *No clearance through military airspace was requested and no weather report was obtained.* Just before take-off, the pilot was informed that he was to route via the Hartebeespoort VOR and on to the Potgietersrus non-directional beacon (NDB).

> **VOR** stands for 'very high frequency omnidirectional radio'. A VOR is a ground-based radio transmitter that sends signals in 360 radials. Some of these radials define airways, but pilots can track any radial to fly a specific path over the gound. VORs operate on frequencies between 108.0 to 177.95 MHz in the VHF band.

The aircraft took off at 04h58 and was handed over to Johannesburg area control. At 05h01 the pilot contacted Johannesburg area control with a request for direct clearance. This was denied as it would have taken him through military airspace. At 05h04 the pilot requested to change to Pietersburg control to ask for direct clearance to Tzaneen. This was also denied with the remark that only SAAF headquarters could give the clearance for a direct flight.

The pilot went ahead anyway and contacted Pietersburg control on his number 2 radio. When the Pietersburg controller heard that the flight did not have military clearance, he directed the pilot to route via the Pietersburg NDB. He was instructed to be below 105 000 feet when 30 miles from the Pietersburg NDB.

Eight miles from the Potgietersrus NDB the pilot received descent clearance from area control and was cleared to contact Pietersburg control. Clearance to 11 000 feet was given. Three miles from the Pietersburg NDB the pilot reported passing 17 000 feet and requested an altimeter setting (QNH). He was told to report at 11 000 feet and the QNH was given as 1027. As the aircraft approached 11 000 feet the pilot was cleared to contact Lowveld control.

> An **altimeter** is an instrument that measures an aeroplane's height above sea level.

A broken-up transmission was received from the aircraft in which the controller was able to read the aircraft's destination. The pilot was directed to contact the controller when 10 miles from Tzaneen. No further transmissions were received.

A witness standing close to the crash site saw the aircraft briefly when it emerged from very dense clouds. He heard what appeared to be a decrease in power and saw the aircraft lose height. The aircraft started to climb again before he lost sight of it. Approximately 5 seconds later he heard the aircraft crash. Other witnesses spoke of very low cloud or mist in the area.

The aircraft crashed into a Haenertsburg pine plantation at about 05h23. This particular pine plantation consisted of mature trees with an average height of 36 metres and a base girth of two metres. Such was their buffering effect that the distance from first impact to last impact was only 177 metres. On hitting the ground, the aircraft burst into flames killing all four occupants.

The force of the impact caused the Falcon 10 to break into small pieces, making it very difficult for investigators to establish the cause of the crash.

The fan-jet of the Falcon 10. The wood in the blades indicates that the jets were still under power at the time of the crash.

The accident *could have* been caused by a combination of factors:

- The pilot had been on duty for seven consecutive days. In that time he had flown a total of 21 hours and 40 minutes, although the mandatory 24-hour rest period had been observed. He was also studying for an imminent examination.

- The co-pilot had only 71 flying hours on this particular type of aircraft.

- The co-pilot was using a Jeppesen flight guide, which differs substantially from the Aerad flight guide with which he was familiar. Significantly, in the Aerad guide, the minimum safe altitude (MSA) is displayed in the corner. In the Jeppesen guide the MSA is contained in a circle at the top. On the Tzaneen cloudbreak procedure, the Jeppesen plate MSA circle contains the following: 'MSA not yet officially published'. In small print below the procedure, there is a note which states: 'Initial approach altitude 8 000 – (6 086) or higher MSA'. The greatest spot height on the plate is in the north-western sector, 4 698 feet. Subsequent to the accident, 18 pilots familiar with the Jeppesen flight guide were quizzed on the initial approach altitude for the Tzaneen cloudbreak procedure. Of the 18, two said that they would have descended to 4 500 feet. Of the remainder, only half read the note referring to the initial approach altitude of 8 000 feet. The others decided that 4 500 feet would be too low in view of the spot heights and said that they would use the en route sector MSA which was given as 8 600 feet on the route facilities chart.

- The crew was given a flight level of 10 500 feet when 50 kilometres outbound. This placed them in a hazardous position relative to the high ground west of Tzaneen.

- The captain did not obtain a meteorological report before take-off. He was thus not prepared for the cloudbreak procedure when he flew.

Missing

Date:	18 October 1990
Conditions:	Heavy, overcast and icy
Aircraft Type:	Piper Seneca II
Pilot Age:	29
Flying Hours:	1457
Location:	Icidi Pass, Drakensberg, KwaZulu-Natal

Synopsis
The aircraft took off from Richards Bay on what should have been a routine return flight to Grand Central Airport in Midrand. It crashed into the Drakensberg, killing the pilot and four passengers. The wreckage was recovered only six months later.

Flight details
The events of 18 October 1990 culminated in a mystery that would captivate the public's imagination for six months. The flight that took off from Richards Bay was to end tragically in the Drakensberg.

Many planes are privately owned by corporations or individuals and chartered out to pilots with commercial licences. The Piper Seneca was such a plane.

On 17 October, Pilot X received a phone call from Pilot Y who had flown the aircraft MPU that day. He reported several faults on aircraft MPU, including three which were to play a major role in this accident:

- The weather radar was unserviceable.

- The automatic direction finder (ADF) was intermittent.

- The two VOR displays differed by approximately 15 degrees. No 1 VOR was operating effectively, and was therefore serviceable.

Pilot Y told Pilot X that in his opinion, the aircraft should not be flown in instrument meteorological conditions until the defects had been rectified.

Pilot X was flying four businessmen from Grand Central Airport in

Midrand to Richards Bay and back again on the same day. Under normal circumstances, this would not have presented any problems. He was an experienced pilot having served in the SAAF where he flew Mirage jetfighters. However, the conditions that day were to prove particularly challenging.

Let's take a step back now and retrace the aircraft's flight, starting with its departure from Grand Central earlier in the day. On approaching Richards Bay the pilot had some difficulty receiving the Richards Bay non-directional beacon (NDB). The aircraft had refuelled in Richards Bay. Prior to his return flight, the pilot phoned Durban International Airport for the latest weather report.

The flight plan indicated a departure from Richards Bay at 14h00, a cruise speed of 290 kilometres per hour and a flight level of 12 000 feet. It was to be a direct flight to Vrede NDB, then to Heidelberg and on to Grand Central with an alternate route to Lanseria. No search and rescue action was requested by the pilot.

Before departure, the pilot arranged with the pilot of a larger commercial aircraft (MXH) which was departing for Johannesburg that day, to give him a weather update en route. This would compensate for his inoperative weather radar. The pilot of MXH overheard one of the passengers on MPU, (Pilot X's aircraft) ask several times why they were not leaving while many other flights were. This put pressure on the pilot to depart as soon as possible.

The aircraft departed at 13h40. At 13h57, MPU reported at 8 000 feet, saying that he was not receiving the Ulundi distance measuring equipment (DME). At 13h58 the aircraft was approaching 10 000 feet and the pilot advised Durban air traffic control that he would not be tracking over-head the Vrede beacon, but south-west of it. This is how the events unfolded:

- At 14h00, MPU requested amendment to his flight plan to remain at 10 000 feet instead of climbing to level 12 000 feet. Approval was given. This proved to be fatal for the occupants.

- At 14h03, Pilot X called Durban ATC, advising that he was not receiving the Ulundi beacon or DME. (This beacon was out of order for a short time due to a power failure.)

- At 14h31, Durban received a message from MPU that he was coming up to the flight information region (FIR) boundary at Newcastle. This message was relayed through MAG 502 (another larger commercial aircraft). Durban asked MAG 502 to inform MPU to contact Johannesburg International on 119.5 MHz, who confirmed the request.

- At 14h38, MPU confirmed he was coming up to the FIR outbound and that he still had no contact with JHB International.

No further transmissions from MPU were received by either Durban or Johannesburg International.

An ALERFA (a second alert for an aircraft suspected missing) was issued at 20h51 that night, and upgraded to DETRESFA (the third alert phase) the following morning at 05h15.

Despite an extensive search lasting over three months, no trace of MPU could be found. It was six months before the wreckage of the aircraft and the remains of the passengers were found by a hiker at Icidi Pass in the Drakensberg.

The following is a transcript of the communication between Durban ATC, MPU (Pilot X's aircraft), MXH and MAG 502:

13:49.00	**MPU:**	MXH this is MPU 119.9. Do you read?
	MXH:	Affirmative, go ahead.
	MPU:	We have no comms with Durban, maybe you could relay if you don't mind, airborne out of Richards Bay 1 340 through 5 000 feet for level 120.
	MXH:	Say again your call sign.
	MPU:	MPU.
	DBN:	MPU Durban is reading you.
13:49.20	**MPU:**	Good afternoon, airborne Richards Bay 1 340 through 5 000 feet for level 120.
13:49.30	**DBN:**	MPU no reported traffic for the

climb flight level 120, report reaching.

MPU: Next call reaching. Could I, uh, have a chat with MXH again?

DBN: Confirm your estimate for the boundary please.

MPU: Will do so in a moment, I just want to find out from MXH the cloud tops.

MXH: We are just out of the cloud tops at 200 and, uh, there is quite a bit of activity we just went through although it didn't show on our radar, uh, a lot of icing and that just about Ulundi.

MPU: Confirm isolated thundershowers.

13:50.20 **MXH:** We're VMC (visual meteorological conditions) on top now and nothing between us and JS (a non-directional beacon situated near Johannesburg International) by the looks of it.

MPU: Thanks very much.

13:51.50 **DBN:** MPU your level climbing?

MPU: Climbing through 60 doing a spiral climb to the west of Richards Bay.

DBN: Say again the level passing.

MPU: Level 60.

13:52.40 **DBN:** MPU report passing UD (Ulundi NDB) and flight level 100.

MPU: Passing UD and level 100, estimate for the FIR is 1 432.

DBN: 1 432, thank you.

13:57.10 **DBN:** MPU, your level climbing?

33

	MPU:	8 000, sir.
	DBN:	Are you picking up the Ulundi DME?
	MPU:	Negative, not yet.
13:58.20	MPU:	Durban, MPU approaching 100. Are we cleared 120?
	DBN:	MPU affirm you will be tracking over VHD at all?
	MPU:	MPU negative overhead VND.
	DBN:	MPU say again you were a bit broken up.
13:59.10	MPU:	MPU won't be tracking overhead VHD sir.
	DBN:	Roger confirm you will be tracking west of VHD?
	MPU:	That is affirmative sir, south-west.
	DBN:	Thank you sir. We have traffic for you K70R a C208 from Hellsgate estimate VHD 1 412, he is at level 120 as well, if you could keep that in mind when routing on uhm abeam of VHD.
13:49.40	MPU:	Roger will do so, MPU.
	MPU:	Durban, MPU confirm you have us on radar?
	DBN:	Negative sir, not at this stage.
14:00.20	MPU:	Durban, MPU are we cleared to remain level 100 instead of 120?
	DBN:	MPU affirm re-cleared level 100.
	MPU:	MPU.
14:03.30	MPU:	Durban, MPU.
	DBN:	MPU go ahead.
	MPU:	Roger we're now steady level 100 and we're not receiving UD nor Ulundi DME, sir.

	DBN:	Thank you report at FIR boundary.
14:03.50	MPU:	Roger next call FIR boundary and, uh, we are squawking 2 000 in case you see something.
14:04.00	DBN:	Unfortunately we are not secondary radar equipped and can't pick up your squawk and at this stage I've got no targets on target.
	MPU:	Roger, roger thank you very much.

At this stage in the transmission, messages from MPU were relayed to and via MAG 502, a large commercial aircraft.

14:30.30	MAG 502:	MAG 502 TOD for FAPM leaving level 150. Go ahead. (No transmissions heard and no call sign).
14:31.20	502:	Durban from Link 502 for MPU, he is just coming up short the FIR boundary, Newcastle.
	DBN:	Is he at the FIR boundary?
	502:	Ja.
	DBN:	Contact Jan Smuts 119.5 please.
	502:	Jan Smuts 119.5, PU.
14:38.10	MPU:	Durban information MPU.
	DBN:	MPU.
14:38.20	MPU:	Roger sir, we're coming up to the FIR outbound and still no contact with Jan Smuts 119.5.
	DBN:	Okay if in range 128.3, try 128.3 over.
14:38.30	MPU:	Roger 128.3, thank you.

There were no further transmissions after this.

The aircraft flew into Icidi Pass and hit a vertical cliff at 9 800 feet. Just 400 feet higher and the pilot would have made it over the Drakensberg. According to the position of the wreckage, the aircraft was 150 kilometres south of the direct track from Richards Bay to the Vrede NDB.

From the wreckage location, it was established that the pilot was 19 degrees to the left of his intended track.

The estimated flight time was 1 hour and 10 minutes, which meant the aircraft crashed into the Drakensberg at about 14h50, shortly after the last transmission to Durban ATC.

A number of factors *could* have contributed to the crash:

- The pilot was under pressure to take off.
- The automatic direction finder was not working properly.
- The radio transmission seemed to be intermittent.
- The pilot was apprehensive about the aircraft's serviceability.
- Due to weather fronts the pilot went further south and west than originally intended.
- The pilot asked to stay at 10 000 feet instead of going to 12 000 feet as originally planned – possibly because he feared icing?

Tiger's Tail

Date:	16 October 1991
Conditions:	Heavy and overcast
Aircraft Type:	Cessna 210L
Pilot Age:	26
Flying Hours:	1 430
Location:	Tiger's Kloof (east of Vryheid), KwaZulu-Natal

Synopsis

On a return flight from Richards Bay to Piet Retief, the pilot of a Cessna was flying visual flight rules (VFR) in bad weather, where the cloud base was dangerously low. The aircraft flew level wings into the mountain, killing everyone on board.

Flight history

This particular accident occurred in weather that was not conducive to VFR flying. At the crash site there was considerably low cloud with visibility of less than 100 feet.

The flight originated in Nelspruit and proceeded to Piet Retief, where another passenger was picked up. It then flew to Pietermaritzburg where the aircraft was refuelled and continued to Richards Bay. On board were employees of the Lowveld Fire Protection Services.

Just before the return flight, a security officer at Richards Bay Airport overheard the pilot tell one of his passengers that he would try to get into Piet Retief. He did not file a flight plan, nor did he contact Durban's weather bureau to find out about the conditions en route. Had he done so, he would have been told that VFR flying was not recommended for the Durban Flight Information Region.

Take-off from Richards Bay was at 14h09 which gave estimated times for the flight reaching the Johannesburg FIR boundary at 14h51 and Piet Retief at 15h00. The pilot relayed the messages through MRV, stating that he was at 1 500 feet, but did not give flight conditions. It seems as if he was flying VFR by keeping below cloud level. A witness testified that he saw the aircraft flying along the valley about 300 feet from the ground and just below the clouds. Cloud visibility was reduced to 100 feet in places. This forced the pilot to climb into the clouds, and it was shortly after this that the witness heard an explosion. Another witness reported hearing a sudden increase in engine power followed by the thud of impact.

The right wing of the aircraft collided with a tree near to the top of a sheer cliff. Having lost a large part of the right wing, the aircraft rolled to the right. Twenty metres past the initial tree impact, the right wing hit the ground. Sixty metres on, the left wing gouged a large furrow in the ground. Two metres further on, the top of the cabin impacted with the ground. The wreckage trail indicated a heading of between 350 and 355 degrees when it collided with the tree.

After striking the ground inverted, the aircraft bounced and fell into

This is the area into which MW1 flew – VFR with low clouds.

the valley about 250 metres from the cabin roof impact point. Fuel in the reservoir tank ignited and burnt the entire cabin section. Everyone on board burnt to death.

The pilot had a portable GPS which was found out of its case but it could not be determined if it had been in use.

The valley the pilot flew along terminated in a buttress that caused the valley to narrow rapidly. There was no radio navigation facility at Piet Retief and because of poor weather conditions, the pilot's only option was to attempt a visual flight below the clouds. Even though the aircraft was 9 kilometres north of its intended course, it was on a good track (from the accident site to Piet Retief), so the pilot was not lost.

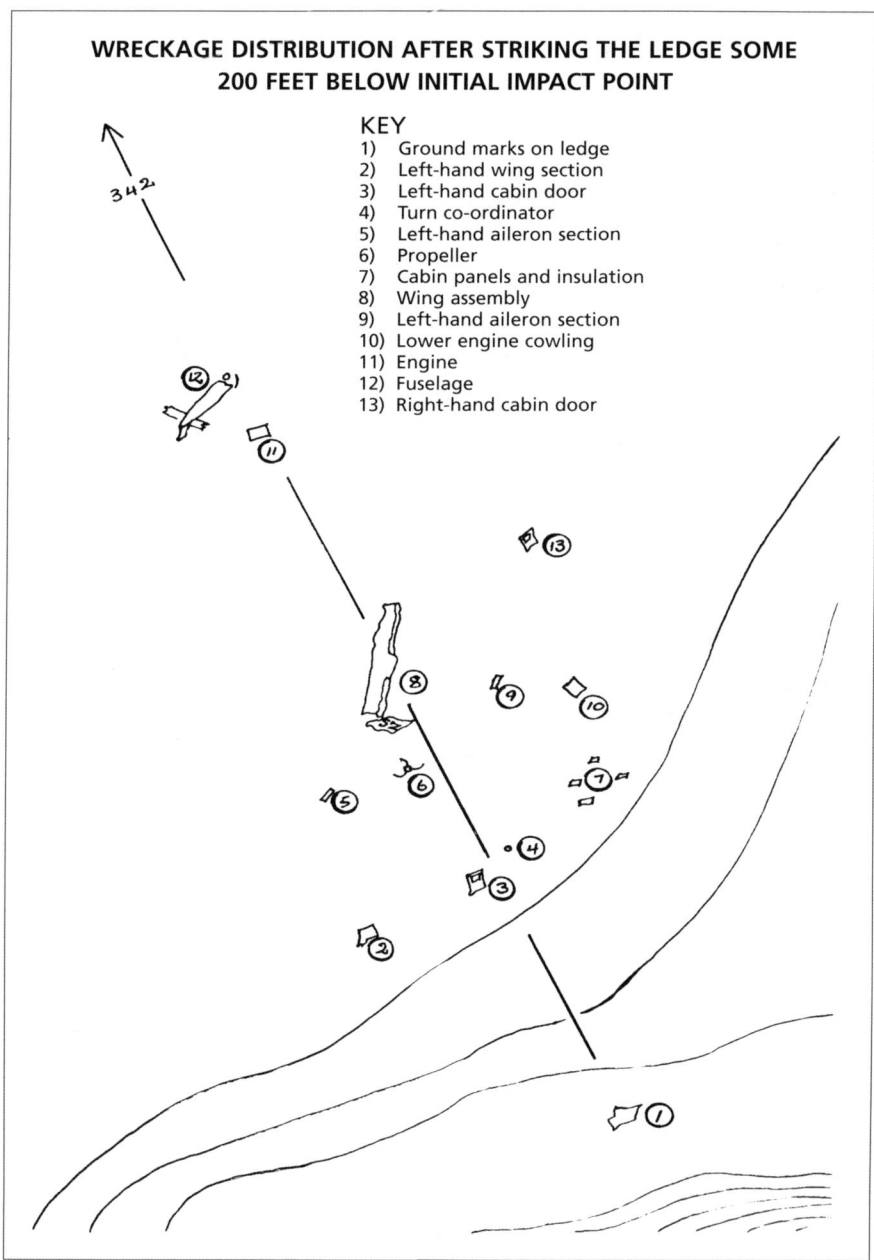

**WRECKAGE DISTRIBUTION AFTER STRIKING THE LEDGE SOME
200 FEET BELOW INITIAL IMPACT POINT**

KEY
1) Ground marks on ledge
2) Left-hand wing section
3) Left-hand cabin door
4) Turn co-ordinator
5) Left-hand aileron section
6) Propeller
7) Cabin panels and insulation
8) Wing assembly
9) Left-hand aileron section
10) Lower engine cowling
11) Engine
12) Fuselage
13) Right-hand cabin door

*Above is the debris map, showing the edge of the cliff to the ledge
where the Cessna eventually stopped.*

Cloudbreak procedures

Date:	5 November 1994, approximately 06h40
Conditions:	Heavily overcast
Aircraft Type:	Piper PA 31-350
Pilot Age:	32
Flying Hours:	6 875
Location:	Trichardtsdal, Penge district, Mpumalanga

Synopsis

This flight from Lanseria to Makalali in Mpumalanga ended in tragedy in remote mountains in the Penge area. The pilot appeared to be flying at low altitude into mountains shrouded in cloud, making flying according to visual flight rules (VFR) very risky.

Flight details

The Piper departed from Lanseria at 05h45, bound for Makalali Game Reserve in the Penge district. The pilot estimated arrival at approximately 06h55. He contacted the Lowveld airspace control sector (LASS) at 06h24 and informed them he was at the flight level of 9 500 feet.

At 06h34 the pilot reported that he was commencing his descent and requested the weather for Hoedspruit. LASS were only able to give him the altimeter setting for Malelane. The pilot then requested details from the pilot of NBO who was flying in the area. He was told by NBO that the cloud base was 5 500 feet above sea level (ASL).

Witnesses in the Penge area stated that during the early part of the morning, the mountains in the area had been shrouded in cloud. The accident occurred between 06h41 and 06h55. The aircraft struck the mountains 29 kilometres short of Makalali airstrip, on a direct track between Lanseria and Makalali. It was ascertained that the aircraft was flying at a height of 5 700 feet ASL when it crashed.

Initially, it seemed as though this was just a case of the pilot being in the wrong place at the wrong time. However, evidence found at the crash site pointed to other possible reasons for the crash. Notes found in the wreckage suggested that a well-prepared but illegal GPS cloud-break procedure may have been used in this flight.

A **cloudbreak** is a procedure to descend through cloud. Each airport has a different cloudbreak procedure, which is described on an approach plate. The **approach plate** specifies the height, direction and method for each cloud-break procedure.

The following quote from a CAA report explains briefly what consti-tutes an approved GPS cloudbreak:

An approved, proper GPS cloudbreak will consist of two or three waypoints, a final approach course fix, a final approach fix and a missed approach point.

The illegal procedure had only one waypoint, no missed approach point and lacked the required 5 nautical mile buffer zone between the high ground and the commencement of the final descent.

In fact, the use of GPS, even for non-precision instrument approaches, is not approved by the CAA in South Africa.

If the illegal approach was followed this does not mean that it was the cause of the accident but it could have contributed to the final outcome. An equally plausible explanation could be that the pilot trans-posed the minimum approach height of 7 500 feet with 5 700 feet. (This was the height he had descended to and was maintaining when the air-craft struck the mountains.)

The wreckage of MXJ lies in a tangled heap in a ravine in the Penge mountains.

41

Spot the gap!

Date:	7 March 1994, between 11h35 and 11h40
Conditions:	Overcast
Aircraft Type:	Piper Seneca PA34-200T
Pilot Age:	51
Flying Hours:	410
Location:	Oudtshoorn, Western Cape

Synopsis
A Piper Seneca crashed while attempting a landing at Oudtshoorn Airport in overcast conditions. A passenger was killed and the pilot badly injured.

Flight details
The flight plan filed by the pilot indicated that it was to be a visual flight rules flight. Estimated flying time was 3 hours and 20 minutes and ETA at George was 11h35. Because he anticipated heavy weather over George, the pilot designated Oudtshoorn as the alternate airport.

At 11h10, George air traffic control (ATC) contacted the pilot (who was non-instrument-rated) and informed him that the weather in George had deteriorated to below VFR limits and that he should divert to Oudtshoorn. The pilot indicated to ATC that Oudtshoorn's weather was also below VFR limits. It was suggested that he divert to Port Elizabeth.

The diversion was accepted but shortly thereafter the pilot advised that he had Oudtshoorn in sight and that he would descend. He was instructed to remain VFR. The pilot then requested runway information at Oudtshoorn and acknowledged the information passed to him. Soon afterwards, the aircraft crashed.

The Piper Seneca lies broken and battered in the bush near the Oudtshoorn military camp.

Following is the transcript of the communication between LPG and George ATC:

LPG: George tower this is LPG.

GEO: LPG good afternoon, go ahead.

LPG: Ah, LPG is approximately 43 miles – could you tell us your weather conditions please?

GEO: We are IMC, cloud base is approximately 200 feet. We are overcast, visibility is approximately 4000 m. Suggest you divert to Oudtshoorn, your present flying conditions?

LPG: Ah, its starting to close up now, ah, will divert to Oudtshoorn. I'll just have to

43

work out another flight – LPG.

GEO: Roger LPG, keep me advised of your intentions.

LPG: Will keep you informed.

GEO: What is your DME from George now?

LPG: We are, ah, 41 miles out.

LPG: George LPG.

GEO: LPG.

LPG: We'll divert to Oudtshoorn, any idea of what the weather is like there?

GEO: Standby for weather Oudtshoorn.

LPG: Thanks very much – LPG.

GEO: Your flying conditions now?

LPG: Ah, seems to be closing up now, quite badly.

GEO: Roger PG, confirm you are VMC below cloud?

LPG: Ah, we are not below cloud. We're in a little bit of brokenness here and I'm scared to go down.

GEO: Are you familiar with the Willowmore area?

LPG: Am I familiar with?

GEO: Willowmore.

LPG: I don't quite get what you are asking me.

GEO: Okay, standby a second.

LPG: Okay.

GEO: LPG – George verify your endurance.

GEO: Your position now?

LPG: LPG go ahead.

GEO: LPG your DME now?

LPG: Ah, well we are first heading like I said. We're going to Oudtshoorn, and ah, we're approximately 15 miles from Oudtshoorn, but the weather doesn't look good here either.

GEO: What does the weather look like north of the Swartberg mountains?

LPG: Well, at the moment we've got clouds all around us.

GEO: Keep me advised of intentions, if you cannot get into Oudtshoorn, I suggest you try Prince Albert.

LPG: Okay we'll keep you advised — LPG.

GEO: LPG report your flying conditions now.

LPG: They're still bad.

GEO: And LPG, could you just confirm your fuel endurance left on board?

LPG: Fuel endurance 40 gallons.

GEO: Forty gallons. How many hours or minutes is that?

LPG: Ah, I'll give you the times now. It will be approximately two hours.

GEO: Roger LPG.

LPG: Okay George — LPG.

LPG: LPG George, do you have any coordinates for Prince Albert?

GEO: Roger, the last report from Prince Albert was that it's raining there and the runway is waterlogged.

LPG:	I see. Any other suggestions?
GEO:	The best bet would be to divert to Port Elizabeth.
LPG:	Port Elizabeth? Any idea what the weather is like there?
GEO:	Okay they've got ... cloud at 2 000 feet broken at 9 000, temperature 26, dew point 18 and visibility is good.
LPG:	I see, okay we'll divert to Port Elizabeth then – LPG.
GEO:	Okay LPG your estimates for boundary and Port Elizabeth?
LPG:	LPG.
LPG:	George LPG.
GEO:	LPG George.
LPG:	We're overhead Oudtshoorn, we have the town visual and we're going to descend a little bit and then I'll see.
GEO:	LPG remain VFR, advise further intentions.
LPG:	Will advise you and will stay VFR – LPG.
GEO:	LPG – your position now?
LPG:	We're descending now ... tell me ... the runway is on the...
GEO:	Okay LPG the runway at Outshoorn is 04/22 and its 1 700 m long.
LPG:	Okay – LPG.
GEO:	LPG we will probably be losing contact shortly, your altitude passing?
GEO:	LPG your altitude?

No further transmissions were received from LPG.

It is likely that the pilot was flying on autopilot under instrument meteorological conditions (even though he was not instrument-rated) and on seeing Oudtshoorn through a cloudbreak, decided to descend. His logbook showed that he had never flown this route before. Although he had designated Oudtshoorn as the alternative aerodrome he had not prepared himself for a possible diversion. His failure to accept George ATC's advice to divert to Port Elizabeth proved to be disastrous.

Judging from photographs the aircraft appeared to have crashed into the slope of a mountain. This accident resulted in the death of a passenger and serious injury to the pilot.

Fatigue

Date:	20 November, 1999, 18h10
Conditions:	Foggy with cloud patches
Aircraft Type:	Beech B55
Pilot Age:	46
Flying Hours:	793
Location:	Vermaakskop, (mountains near Uitenhage) Eastern Cape

Synopsis
The pilot was on the last leg of a flight between Namibia and Port Elizabeth when he crashed into a mountain ridge.

Flight details
After spending six weeks in Namibia completing a business contract, the pilot, who owned an industrial installation company, was flying home for a holiday. He had refuelled in Upington from where he phoned his fiancée at 15h40 to say he was leaving shortly. She confirmed that the weather in Port Elizabeth was good.

Following is the transcript of the communication between the pilot (KLE) and Port Elizabeth air traffic control (ATC):

17:32:57   **ATC:** Kilo Lima Echo, Port Elizabeth, are you on frequency?

17:33:02   **KLE:** PE from Kilo Lima Echo, we read presently nine five miles inbound flight level one one five.

17:33:09   **ATC:** Thank you Kilo Lima Echo. Report passing abeam Charlie Hotel, runway two six is in use at Port Elizabeth. Our QNH one zero one seven. We have few clouds at one thousand five hundred feet and scattered cloud two thousand five hundred feet.

17:33:26   **KLE:** Thank you, we copy that QNH one zero one seven and err - Charlie Hotel next Kilo Lima Echo.

17:33:32   **ATC:** Kilo Lima Echo.

17:43:43   **KLE:** From Kilo Lima Echo requesting descent out of one one five.

17:43:48   **ATC:** Kilo Lima Echo descend VFR, QNH one zero one seven report passing Uitenhage.

17:43:55   **KLE:** Thank you. Cleared for descent and Uitenhage next Kilo Lima Echo.

17:43:59   **ATC:** Kilo Lima Echo.

17:57:27   **ATC:** Kilo Lima Echo, Port Elizabeth.

17:58:40   **ATC:** Kilo Lima Echo, Port Elizabeth.

18:01:16   **ATC:** Kilo Lima Echo, Port Elizabeth.

18:02:39   **ATC:** Kilo Lima Echo, Port Elizabeth, do you read?

18:06:46   **ATC:** Kilo Lima Echo, Port Elizabeth, one two four seven, do you read?

18:08:37   **ATC:** Kilo Lima Echo, Port Elizabeth, one two four seven, do you read?

KLE impacted the ground 15 metres from the top of a ridge at an altitude of 2 625 feet on a magnetic heading of 165 degrees. The aircraft was on track to Port Elizabeth.

So what did the investigators find?

According to the ground strike marks, the aircraft appeared to have flown into the ridge 'level wings'. This indicates that the pilot thought that he was higher than he was.

Secondly, the main wreckage was located 209 metres from the initial point of impact and the engines were located at the bottom of the valley, some 418 metres from the initial impact point.

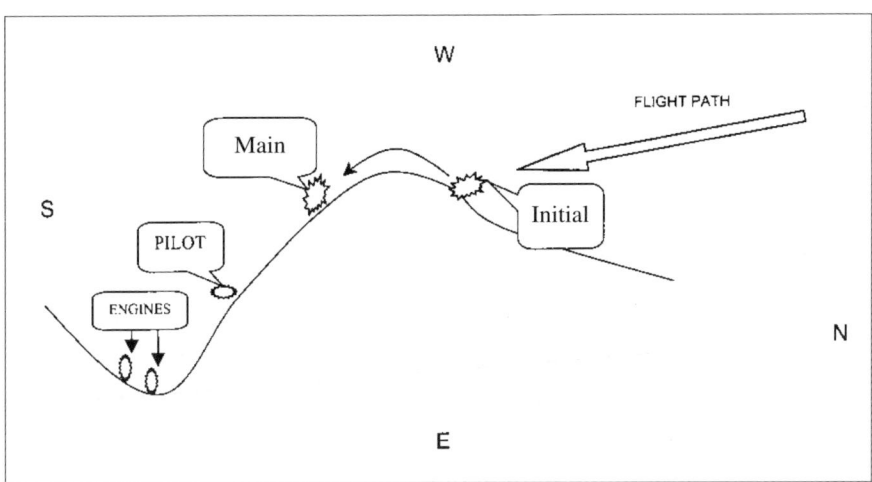

A diagram of the crash impact and flight path angle.

This would suggest that the aircraft was at cruise speed and that the pilot did not suspect that he was in any danger.

The altimeter was recovered and found to be set at 1 013 hecto-pascals (hPa). The QNH supplied by ATC was 1 017 hPa. This would have resulted in an over-read of 120 feet. This would still have been very close to the ground but would in all probability have missed the ground.

This accident was almost certainly a case of controlled flight into terrain. The pilot was more than likely fatigued as a result of six weeks of hard work and the long flight back from Namibia. One slight miscalculation or a descent made a fraction too soon can be all it takes to cause a fatal accident.

Fast and low

Date:	12 December 1999
Conditions:	Heavy fog/low cloud
Aircraft Type:	Piper PA32-300
Pilot Age:	45
Flying Hours:	705
Location:	Plettenberg Bay, Western Cape

Synopsis
A flight from Kimberley to Plettenberg Bay ended in tragedy for three people when the aircraft hit power line conductors and crashed.

Flight details
The flight took off from Kimberley at about 05h00. It was uneventful until outside Plettenberg Bay, when fog and low cloud started to drift in. Visibility therefore was very bad.

The Plettenberg Bay Airport manager heard the pilot of NAR calling that he was inbound and descending through 1 500 feet. He transmitted back saying that conditions were very bad and that he should not attempt to land at Plettenberg Bay.

Following is the transcript of the communication between NAR and George ATC:

07h00	**NAR:**	George, NAR are you there?
	ATC:	NAR go ahead.
	NAR:	Thank you Sir. We are Cherokee 6 inbound from Kimberley to Plettenberg Bay, the present time, uh, we are on a track 216 degrees range 51 miles from Plett. We estimate Plettenberg Bay in 20 minutes, flight level 085 and top of descent at the present time, NAR.

ATC:	NAR, no reported traffic for Plettenberg Bay, continue on the VFR frequency.
NAR:	Thank you sir, we're going to 124.8 for Plett, NAR.
NAR:	George, NAR, could you give me your QNH please.
ATC:	Last caller, George, say again please.
NAR:	George, this is NAR, could you give me your QNH please NAR?
ATC:	QNH George 1017.
NAR:	Thank you sir, NAR.
ATC:	The wind is 120°/10 knots.
NAR:	Thank you sir, appreciate that help, NAR.

RECORDING 118.9 STOPPED AT 07h00.

RECORDING 124.8 STARTING AT 06h50

NAR:	General area traffic in the Plettenberg area, this is NAR presently 7 miles from yours descending for 4 000 feet QNH 1019 any conflicting traffic Plettenberg Bay, NAR.

The following transmissions by the pilot were made to an unknown person/pilot.

NAR:	Sorry, are you calling for me, NAR?
NAR:	OK Paul, we're 2 miles out and at 3 000 feet, what do you want, talk quickly.
NAR:	OK, we are going into cloud now.

51

I'm going into the cloud right now. I am one mile from the airport. I should be at the airport in 34 seconds, we are entering cloud now at 3 500 feet and see you in a couple of seconds, NAR, here we go.

NAR: Go ahead, NAR.

Other pilot: Please report the altitude of yourself now please!

Seconds after this a witness saw NAR emerge low from the clouds in a very steep right-hand bank. The aircraft had just levelled out when it hit the high voltage conductors about 50 metres from where the witness was standing.

It pitched up and disappeared into the fog. It then crashed into the ground at a relatively high speed, nose-down and cartwheeled twice before coming to a stop 38 metres away. This occurred about 1 kilometre from where it first hit the high voltage conductors.

The charred remains of a Piper lie 1,8 km from Plettenberg Bay Airport.

As with most airports, there is a cloudbreak procedure for Plettenberg Bay. The minimum approach altitude is given as 6 300 feet. The obstacle clearance height (OCH) for the approach is given as 844 feet and thus the minimum descent altitude will be 1 309 feet (aerodrome altitude is 465 feet). All this assumes that the pilot is instrument-rated. The pilot of NAR was not. He was not entitled to do a cloudbreak procedure yet he descended to 30 feet – suicidally low!

Did the pilot allow himself to be distracted? The fact that he told the unknown person on the ground that he should speak quickly shows that he was running out of time and needed to concentrate on the task at hand. He obviously intended to fly beneath the clouds for a landing at Plettenberg Bay Airport, but misjudged the cloud base. The true story died along with the pilot.

Inexperience and impatience

Date:	3 April 1999, 09h14
Conditions:	Heavy, low overcast
Aircraft Type:	Beech A36 (Bonanza)
Pilot Age:	42
Flying Hours:	114
Location:	Near Port Elizabeth, Eastern Cape

Synopsis
On a flight from Klerksdorp in Gauteng to Paradise Beach near Jeffrey's Bay, the pilot was diverted to Darlington Dam because of bad weather. While trying to fly below the cloud base, the aircraft impacted the ground about 13 kilometres from Port Elizabeth.

Flight details
The pilot left Klerksdorp early in the morning on a private visual flight rules (VFR) flight and was due to meet his family for an Easter weekend holiday at Paradise Beach near Jeffrey's Bay.

He had flown this route twice before, both times with an instructor. On both flights the weather played a role. On the first flight they encountered airframe icing and descended and diverted to Bloemfontein. The second time the weather report obtained before their departure from Paradise Beach did not allow for VFR, so the flight was postponed till the following day. Both times, the decision to divert or not to fly was made by the instructor.

On the day of the fatal flight, the pilot began to encounter broken cloud and was called by Port Elizabeth air traffic control (ATC), who advised him to divert and land at Darlington Dam, which is about 111 kilometres north of Paradise Beach. The weather along the coast had deteriorated to instrument meteorological conditions (IMC). The pilot had some difficulty in finding the landing spot but eventually landed safely. He waited at Darlington Dam for about one and a half hours and then took off again.

Why the pilot did this, we will never know, but it was to prove fatal for him. At 09h04 he told ATC that he was running out of fuel and that the cloud cover was very low. ATC suggested that he fly towards Uitenhage, but he was reluctant to enter cloud. At 09h05 he stated that he was looking for a place to break through the clouds but that he thought he was going to hit a mountain.

The transcript of the communication between the pilot and George ATC reflects his growing desperation. ATC tried to get him to backtrack to Darlington Dam, but he felt that he could not get there on his remaining fuel. The communications between ATC and the pilot were very tense but suddenly, a break in cloud appeared. The pilot advised that he was going to try to get through the cloud and fly visual flight rules (VFR). He gave his altitude as 1 500 feet and said that there was not much high ground around him. He then said that he would have to fly at 1 000 feet or lower to stay below the clouds.

At this stage, ATC told him that he was about 7 miles from Port Elizabeth Airport and that he should route to join with the airport circuit. ATC called several times but got no answer. They asked a commercial flight to try to pick up PEW on its radar screen. The pilots on this flight responded that they had seen him for a second, but that he had quickly disappeared from their screen.

There were no further transmissions from PEW after 09h14. The aircraft hit the ground approximately 13 kilometres north-west of Port

Elizabeth at 650 feet above mean sea level. The crash left a scar of 150 metres and the investigators concluded that the aircraft was in a slight descent, with a shallow bank angle to the right on a heading of 180 degrees magnetic. This was obviously a controlled flight into terrain (CFIT) accident. The pilot believed that he was flying with enough clearance from the ground. Keen to get to his destination that day, he made several crucial errors.

Let's look at the facts of the flight:

- The pilot was not instrument rated, so was limited to flying VFR at all times. Instead of turning back at the first signs of bad weather, he flew on.

- Although the aircraft was fitted with a transponder equipped with Mode C, it was not switched to that mode. Had Mode C been switched on, ATC would have been able to readily identify PEW. They would have been able to tell his altitude and warn him of any danger.

- The pilot kept referring to a lack of fuel and mentioned having a quarter of a tank left in each wing. This would have bought him a full hour of flying time, more than enough time to get back to Darlington Dam, which was a mere twenty minutes away.

- High stress conditions can often affect one's judgement.

The ATC aircraft transcript is chilling evidence of what went wrong on this flight. And yet it could have been so different. Had the pilot stayed at Darlington Dam or diverted to Graaff Reinet, he would be alive today. We will never quite know why he changed his mind only to fly to a rendezvous with death.

CHAPTER 5

Aerobatics

PERFORMING AEROBATICS IN a non-aerobatic aircraft is not only dangerous and irresponsible – it is also illegal. The CAA files cite many examples of pilots who tried to perform aerobatics in aircraft that were not custom made to turn tricks.

Sadly too, many of the following cases involved younger pilots, who had boasted about their intended feats to friends.

Tragically, all of these cases ended in death or serious injury.

Dangerous manoeuvres

Date:	10 March 1993, 14h40
Aircraft Type:	Piper PA34-200T
Pilot Age:	25
Flying Hours:	380.5
Location:	Westonaria, Gauteng

Synopsis
A joyride ended in tragedy for three people when the aircraft they were flying in suddenly broke up mid-air. How did this happen? Did the pilot make a fatal error in carrying out an illegal aerobatic manoeuvre? Or was the aircraft defective?

Flight details
The aircraft took off from Rand Airport at 14h00, carrying the pilot and two passengers. It was bound for the Johannesburg general flying area.

About 40 minutes later, a witness reported seeing the aircraft with level wings, but a slight (10 to 15 degrees) nose-up attitude. Seconds later it had broken up mid-air and was plummeting to the ground. Two other witnesses also told how they saw pieces of aircraft falling away and the fuselage hitting the ground, eventually ending up in a storm-water drain.

Despite extensive enquiries, nobody had actually seen the pilot's manoeuvres prior to the breaking up of the aircraft.

What information do we have about the pilot?
The pilot was 25 years old. He held a valid license with a total of 379 flying hours to his credit. He had recorded 12 hours on this particular aircraft. According to his logbook, he had flown the PA34-200T for an hour the previous day, accompanied by an instructor. It was also noted that he had undergone 6 hours and 40 minutes of aerobatic instruction in a Pitts S2-B.

To be able to hire the PA34, the operator required that he pass a flight check with an instructor. The check was carried out the day before the accident. Shortly after this, the operator received a report from the instructor that the pilot had performed a barrel roll during the flight check. This claim had been substantiated by two other witnesses. The instructor had not given permission for such a manoeuvre. The pilot had also told his instructor that he owned a Bosbok aircraft in which he carried out aerobatic manoeuvres.

A Bosbok, like the PA34-200T, is an aircraft in which all intentional aerobatic manoeuvres, including spins, are prohibited.

On examination, the main wing spars were found to have failed in the positive upward mode. This was because the pilot pulled the elevator to the full-up position while flying in excess of the maximum recommended speed.

> The **main wing spar** is a rigid beam that connects the two wings providing strength and support.

Did the pilot carry out a manoeuvre that caused him to lose control of the aircraft? Did he then pull up too sharply while exceeding the maximum speed of the aircraft? We will never know exactly what happened, but clearly the aircraft was stressed beyond its capabilities.

CAA investigators examine the wreckage of the Piper lying in a storm-water drain.

Aerobatic crop spraying

Date:	30 October 1999, 07h50
Conditions:	Fine with no wind
Aircraft Type:	Ayres S2R-T34
Pilot Age:	41
Flying Hours:	7 490
Location:	Near Reitz, Free State

Synopsis
A crop sprayer pilot did a low level pass over a resort and then appeared to perform a roll. The aircraft did not right itself in time and impacted with the ground, killing the pilot instantly.

Flight details
After completing a crop spraying job at a local farm, the pilot of an Ayres flew over the Bietjiewater holiday resort at 1 000 feet AGL. He then did what looked like a roll, according to witnesses who were 500 metres from the impact point.

At such a low altitude, recovery from such a manoeuvre would have been difficult for most aircraft. Due to its heaviness and relatively slow speed, aerobatic manoeuvres are strictly prohibited on Ayres aircraft. The crop sprayer struck the ground at speed and exploded. The pilot was killed instantly.

According to the pilot's family and friends, he was extremely conscientious and responsible. He was an experienced pilot with many flying hours, so why did he carry out such a risky manoeuvre?

The following is a quote directly from the CAA records on this accident: 'Part 91.06.32 of the Civil Aviation Regulations of 1997(CAR) requires that pilots fly more than 1 000 feet above the highest obstacle, within a radius of 2 000 feet, over an open-air assembly of persons.' Although accident investigators looked for evidence that a mechanical breakage caused the accident, none was found.

The Ayres' S2R-T34 flightpath over the trees to impact point.

The Ayres crop sprayer lies inverted and burnt out.

A joyride ends in tragedy

Date:	21 March 1992
Conditions:	Fine, with a 10- to 15-knot wind
Aircraft Type:	Piper PA28-140
Pilot Age:	19
Flying Hours:	19
Location:	Eshowe, KwaZulu-Natal

Synopsis

Having recently obtained his private license, a young pilot took three of his friends for a joyride. While performing a series of illegally low manoeuvres, the aircraft crashed into a road in a suburb of Eshowe.

Flight details

The pilot had completed his private pilot's license (PPL) on 27 February 1992, just four weeks before the fatal flight. His flying record suggests that he was 'quick to learn' but ten days earlier an instructor had stated that 'the learner pilot lacks personal discipline in his flying'.

On the fateful day, he hired the Piper from the Natal Flight Centre at Virginia Airport, and flew up to Eshowe in Zululand. A witness saw the aircraft earlier that day flying low and executing steep turns over the built-up area of the town.

The same witness states that he was in his study later in the day when he heard the sound of an aircraft coming from the direction of Eshowe Airport. He saw the aircraft briefly through a break in some trees and recognised it as the Piper he had seen earlier. It was flying at a dangerously low altitude for a suburban area, not more than 150 feet AGL. He lost sight of it momentarily, but when it reappeared, it was still low and beginning a steep left-hand turn. Just then the telephone rang and he went to answer it, still aware of the aircraft engine noise in the background. About 20 to 30 seconds later, he heard the sickening 'thud…thud' of a double impact. Running to the window, he saw a cloud of dust some 150 metres away from his home.

The witness, who was a local surgeon, ran to the crash site to see if he could assist in any way. One of the occupants showed some signs of life, but died shortly afterwards. The other three occupants had died instantly.

Another witness stated that the aircraft flew towards him at an estimated height of 130 feet AGL. The aircraft flew over him and then commenced a right-hand turn. The turn became progressively steeper, the nose dropped to almost 45° in the turn and the left wing struck the ground. It bounced twice and crossed a road before coming to a stop on the driveway of a private residence.

The accident happened in a 145 m² field. At one edge of the field were high-tension wires and trees that were 6 to 8 metres high. The aircraft was turning towards these. It is possible that the pilot tried to tighten his turn to avoid them. In doing so, the aircraft stalled and crashed.

A disturbing aspect of this case is the fact that a Kamikaze bandana, one such as those worn by Japanese pilots during World War Two, was found in the wreckage. It was assumed to belong to the pilot, who had been known to use the call sign 'Tora'.

The scarf allegedly worn by the pilot of the doomed PA28-140, shows a rising sun emblem, similar to those worn by the Japanese pilots in World War Two.

CHAPTER 6

Mid-air collisions

IN SOUTH AFRICA we are fortunate to have had very few cases of mid-air collisions, although there have been a number of 'near misses'.

In 1982 a military aircraft and a private plane collided over Pretoria. Since then the airspace conditions over South Africa have improved drastically, although the skies above Johannesburg remain extremely congested.

One of the most vivid mid-air crashes that I recall occurred on 25 September 1978, in San Diego in the United States. A Boeing 727, coming into land at the city's airport, collided with a Cessna 172 that had strayed into the Boeing's flight path. A total of 137 people were killed. Although the two aircraft were flying in the same direction (they were heading for the airport and were flying reasonably low and slow), the Boeing's air speed was far higher than the Cessna's and it hit the smaller aircraft from behind. It damaged vital aerodynamic components on the wing of the Boeing and caused it to nosedive into the ground.

The fickle finger of fate

Date:	14 July 1982, 16h35
Aircraft Type:	Swearingen SA 226 AT/Piper PA31
Pilots' Age:	28 and 46
Flying Hours:	2 818 and 5 137
Location:	Pretoria, Gauteng

Synopsis
Two aircraft flying from different areas of South Africa collided near Pretoria, killing all occupants on board.

Flight details
Mid-air accidents are a strange phenomenon. Flying a plane is not like driving a car, where you can only choose to swerve left or right. In the air, you can slow down, speed up, move up or down, left or right … the chances of meeting with another aircraft seem negligible.

In this case, one of the aircraft was an SAAF plane on a trip from Grootfontein in Namibia, the other a private plane from Vereeniging bound for Wonderboom Airport, north of Pretoria.

Let's look at the SAAF 16's flight in more detail:

SAAF 16 flew from Grootfontein with two pilots, one air hostess and five passengers. Its destination was Waterkloof Airbase, south of Pretoria. An IFR flight plan filed at Grootfontein predicted a flight level of 15 000 feet, routing over Gaborone and Hartebeespoort Dam VOR to Waterkloof Airbase. The flight should have lasted three hours, at a cruise speed of 400 kilometres per hour. This was how the tragedy unfolded:

- At 16:26:28, the pilot requested Johannesburg International approach control for descent clearance from 15 000 feet to 12 000 feet. This was granted.

- At 16:30:04, the aircraft was re-cleared to 11 000 feet and at 16:31:20 the pilot reported that he was approaching this level. At that time, the transponder at Johannesburg International Airport indicated a flight level of 11 600 feet.

- At 16:31:25, the aircraft was cleared to descend to 8 000 feet.

- At 16:33, the pilot informed Waterkloof air traffic control that they had just passed Hartebeespoort Dam VOR and were 30 kilometres from Waterkloof, descending through 9 000 feet. The pilot was asked if he would accept a visual approach to Waterkloof Airbase. He replied that he would. He was advised that there was no known traffic at Swartkops and that he could continue for a visual approach onto runway 01.

- At 16:34, the pilot read back ATC's instructions and was cleared down to 6 000 feet. He confirmed that he would call back when he reached the cleared altitude.

- At 16:35:12, an emergency locator transmission was heard.

> An **emergency locator** is fitted to certain aircraft and transmits a signal in the event of a crash in order to help rescuers locate the aircraft and possible survivors.

Now let's look at what happened to KTX, a private plane on a return flight from Vereeniging:

The pilot had arranged with ATC to do a practice cloudbreak at Wonderboom Airport during the return flight from Vereeniging. He had booked an instrument rating test for the next day. Cloudbreak procedures are done over Wonderboom at 7 000 feet.

- At 16:26:11 the pilot requested to fly over Lanseria Airport at 6 900 feet. Permission was granted.

- At 16:30:30, the pilot reported the Lanseria non-directional beacon at 6 900 feet and bade farewell to ATC at Lanseria.

- At 16:34:35, the pilot's only transmission to the unmanned Wonderboom tower was 'Wonderboom KTX 120.6'.

Fewer than 30 seconds later, the two aircraft collided. Both aircraft were totally destroyed: there were no survivors.

Investigators found that KTX's left wing had hit SAAF 16's right wing and body. Bear in mind that the aircraft were probably travelling at near to cruising speed, so the combined kinetic energy must have been immense. The pilot of KTX was probably wearing a blind flying hood, so he would have been on the 'blind' side of the incoming SAAF aircraft. The co-pilot, who was flying safety pilot on this flight, could probably not have seen the SAAF aircraft from his seat. As it was dusk it would have been difficult to see another aircraft against the backdrop of lights on the ground.

As for the SAAF 16, the fact that ATC had said that there was 'no known traffic at Swartkops', was probably meant to indicate that there was no known traffic flying from Swartkops Airbase, which SAAF 16 had to cross. Could that call have caused the pilot to become complacent? Is it possible that after the long flight from Grootfontein the pilots let their guard slip just a fraction?

Military and civil investigators examine the wreckage that resulted from the mid-air collision.

CHAPTER 7

Helicopters

IT HAS BEEN said that helicopters are a collection of nuts and bolts that should not fly. These ungainly-looking beasts have been around for decades and have proved their worth in different areas of industry many times over. I am of the opinion that if you haven't flown in a helicopter, you haven't flown!

Unfortunately, these beasts are also more difficult to fly than fixed-winged aircraft. The helicopter pilot has to contend with the up and down of the collective pitch control, the left, right, forward and backward movement of the control stick, as well as using the foot pedals to control the tail rotor, which stops the chopper from spinning on its own axis. The wings of a helicopter spin around at high speed and if they, or the tail rotor, hit an obstacle the aircraft can plunge madly to the ground.

As more and more helicopters are being used, so more accidents are occurring.

Power lines are one of the helicopter pilot's worst enemies. In fact, pilots are trained to fly over pylons as the lines themselves are very difficult to spot.

Unforeseen circumstances

Date:	4 September 1999, 08h05
Conditions:	Fine
Aircraft Type:	Bell 206B
Pilot Age:	38
Flying Hours:	Fixed-wing 595; Rotary-wing 4 167
Location:	Copperton, 70 km from Prieska, Northern Cape

Synopsis
While doing maintenance work, a Bell 206B flew into a cable which was lying diagonally across some power lines. The pilot autorotated and the helicopter crashed on landing.

Flight details
Photographs of the location of the power lines provide key information as to how this accident happened:
The helicopter was flying parallel to the power lines (see arrow) when it struck a cable that was lying across the lines (see line). The cable was

corroded and thus very difficult for the pilot to spot. After striking the cable the pilot immediately carried out an autorotation. However, on landing, the skids slid forward for a few metres before digging into the ground. The fuselage tilted forward and caused the main rotor to hit the ground. The main rotor then hit the tail boom, which was severed.

The Bell's tail boom was severed by the main rotor.

Disorientated

Date:	27 December 1998
Conditions:	Overcast with very low cloud
Aircraft Type:	Robinson R44
Pilot Age:	53
Flying Hours:	375
Location:	Muden, near Greytown, KwaZulu-Natal

Synopsis
While on a flight from Phoenix outside Durban to Muden near Greytown, a Robinson helicopter encountered low clouds and rain. In trying to get through the bad weather the pilot became disorientated and crashed into the side of a mountain.

Flight details

Early on 27 December, the pilot, his brother-in-law and two grandsons decided to visit family in Muden, some 100 kilometres north-west of Durban. The weather was against them from the start, overcast with mist and rain and certainly not conducive to flying visual flight rules. However, thinking he could get through by flying under the clouds, the pilot took off after filling up with fuel at Virginia Airport in Durban.

At approximately 08h30, he contacted his family in Muden, telling them that he was experiencing bad weather with low cloud in the vicinity of Merthley Lake. He said he would try to fly via Keate's Drift, a low-lying area 22 kilometres from Greytown.

At 08h45, a witness on the ground saw the chopper and noted that the engine noise had increased, and that it was leaning slightly to the right. It proceeded along the hill and then did a sharp right turn up the valley, with its nose slightly down. Immediately afterwards it crashed into the side of the mountain. The witness ran to the crash site and found three of the occupants dead at the scene. One of the children was alive but seriously hurt. The witness then ran six kilometres to report the accident.

This case clearly demonstrates how dangerous it can be to fly visual flight rules in bad weather conditions. The fact that the chopper appeared to be leaning to the right could well indicate that the chopper was slightly out of control at this point. Realising this, the pilot may have tried to correct his position, pulling himself right into the mountain.

Why did this family outing have to end in tragedy? Why did the pilot not simply turn around and backtrack or find a suitable landing spot and wait for better weather? We will never know.

The following safety notice (SN-18) was taken from the *Pilot's Operating Handbook*:

> Flying a helicopter in obscured visibility due to fog, snow, low ceiling or even a dark night can be fatal. Helicopters have less inherent stability and much faster roll and pitch rates than airplanes. Loss of the pilot's outside visual references, even for a moment, can result in disorientation, wrong control inputs and an uncontrolled crash. This type of situation is likely to occur when a pilot attempts to fly through a partially obscured area and realises too late that he is losing visibility. He loses control of the helicopter when he attempts a turn to regain visibility but is unable to complete the turn without visual references.

You must take corrective action before visibility is lost! Remember, unlike the airplane, the unique capability of the helicopter allows you to land and use alternate transportation during bad weather, provided you have the good judgement and necessary will power to make the correct decision.

This photo shows the mountainous terrain into which the pilot was flying. Flying in these conditions as a non-instrument rated pilot is extremely difficult, if not impossible.

Gyrocopter

Date:	13 May 2000
Conditions:	Fine
Aircraft Type:	VPM 14 Gyrocopter
Pilot Age:	38
Flying Hours:	281
Location:	Approximately 30 km from George, Western Cape

Synopsis
On a flight between George and Swellendam, a gyrocopter suddenly pitched up and then dived into the ground, killing the pilot.

71

A **gyrocopter** is a hybrid between a fixed-wing aircraft and a helicopter. Its momentum is derived from an engine as well as rotors which give the gyrocopter the lift to take off.

Flight details

The gyrocopter left George Airport at 05h53. It was on course for Swellendam, where it was to take part in an air show the following day. It was flying in loose formation with a microlight when about 30 km from George, it suddenly pitched up, banked to the right and dived into the ground.

The microlight pilot, oblivious to the drama that had just unfolded, continued on his way. On reaching Swellendam, he became concerned about the gyrocopter's whereabouts. An aerial search was initiated and some six and a half hours later, the wreckage of the gyrocopter was spotted on the farm 'Hartenbos'.

This was not the first time this particular aircraft had been in an accident. The gyrocopter was built in 1992, after being bought as a kit. On his first solo flight, the original pilot lost control and crashed. The wreckage was then bought by a businessman in Vredendal who rebuilt it, before selling it in February 1998 to the pilot involved in the fatal crash.

According to the logbook, the aircraft was also involved in an accident on 1 March 1998, when the pilot attempted a short field take-off. He was unable to get airborne due to insufficient rotation speed and the aircraft collided with a fence. This accident was not reported to the CAA.

Again, the aircraft was extensively rebuilt. The list of parts requiring attention read as follows:

- replace rotor control rods
- rebuild rotor control fork
- replace all control system bearings
- replace all rod-end bearings.

On inspecting the crash site investigators found that all controls worked perfectly, except for one of the push rods. It had either broken on impact or come loose during the flight. Extensive testing was done on the remaining push rods and the related rod-end shafts to establish which one of these had failed. Interestingly, the rod-end shaft proved to be the most resilient when subjected to tension tests.

Due to the fact that the left-hand control arm rod-end bolt was missing, two scenarios were considered. One possibility was that the bolt had failed on impact, become dislodged and could not be found. The other possible scenario was that the bolt had become dislodged in flight and had fallen off. Investigators considered the second scenario to be more likely. It appeared that the bolt had not been correctly fitted. This was supported by the fact that there were no scratch marks on the control arm bracket where the rod-end would have been fitted.

A view of the wrecked gyrocopter. One of the rotor blades lies in the foreground.

A photograph of the gyrocopter's control arm, which shows that the push rod was not bolted to the control bracket.

It is believed that when the left-hand control arm's rod-end bolt became dislodged during flight, it became unrestricted, causing the rotor to tilt towards the right. This had caused the airframe to separate prior to impact.

The two eyewitnesses who saw the aircraft bank towards the right and plummet to the ground, further supported this conclusion.

Company rules

Date:	27 August 1999
Conditions:	Fine
Aircraft Type:	Robinson R-44
Pilot Age:	41
Flying Hours:	6 060
Location:	Soweto, Gauteng

Synopsis
The helicopter, equipped with vehicle tracking systems, was assisting in the recovery of a hijacked vehicle. While circling the target, it struck power lines and crashed, injuring the pilot and killing his passenger.

Flight details
The call came at about 10h30 … 'Vehicle hijacked.' The crew of the R-44 scrambled to prepare the Robinson and it was airborne by 10h35. On board were the pilot and an observer.

They departed from Grand Central Airport and headed towards Soweto, on the opposite side of Johannesburg. They soon picked up a signal from the vehicle and started to track it. At the same time, the observer contacted the ground vehicles and directed them towards the hijacked vehicle. While the drama unfolded on the ground below, the Robinson was circling at 6 800 feet AGL. After a short chase, the hijacked vehicle was stopped and the suspects were apprehended.

Shortly afterwards, the pilot heard a loud bang as he started to lose control of the chopper. His immediate reaction was to lower the cyclic pitch and head for a safe landing spot. Despite all his efforts, the chopper hit power lines and crashed.

The pilot had unwittingly strayed well below the prescribed minimum safety height, possibly because he was mesmerised by the activity below and had positioned the chopper close to the power lines. Once the blades had struck, the fate of the chopper was sealed.

75

The tracking company's operating manual states the following with regard to minimum heights:

- All flights to be flown at the minimum heights required by the ANR's (Air Navigation Regulations – 1976) provided that by *day* (the company) has written exemption to 500 ft AGL *while tracking the vehicle*. Time at 500 ft to be kept to a minimum.

- If minimum height cannot be maintained Day or Night due to Met conditions, then the flight must be postponed/cancelled.

- Minimum heights by *day* are: 1 000 ft above the highest obstacle within 2 000 ft.

- Minimum heights by *night* are: 1 500 ft above the highest obstacle within 5 nautical miles.

In the Pilot's Operating Handbook (POH), Section 10, Safety Tips, Safety Notice SN-16, the following is stated:

'Flying into wires, cables, and other objects is by far the number one cause of fatal accidents in helicopters. Pilots must constantly be on the alert for this very real hazard.

- Watch for the towers, you will not see the wires in time.

- Fly directly over the towers when crossing power lines.

- Allow for the smaller, usually invisible, grounding wire(s) which are well above the larger more visible wires.

- Constantly scan the higher terrain on either side of your flight path for towers.

- Always *maintain at least 500 feet AGL* except during take-off and landing. By always flying above 500 feet AGL, you can virtually eliminate the primary cause of fatal accidents.'

The Robinson R-44 lies on its left side after crashing into power lines over Soweto.

The R-44's tail rotor lies a couple of metres from the main wreckage.

CHAPTER 8

Fuel mismanagement

FUEL IS THE lifeblood of any aircraft. So why do so many aircraft run out of fuel? One would think that because pilots' lives depend on having sufficient fuel, levels would be checked fastidiously. Granted, there are situations where marginal flying is unavoidable, such as when a pilot has to divert because of bad weather. What is difficult to understand though is the attitude of those who do not fill up because they 'feel' they have enough. This is an unforgiveable safety risk.

No fuel – no power!

Date:	24 October 1991
Aircraft Type:	Piper PA28R-200
Pilot Age:	33
Flying Hours:	281
Location:	5,4 km south-west of Welkom Airport, Free State

Synopsis

Despite there being enough fuel left in the Piper's right-hand tank, the pilot mismanaged the fuel and the engine stopped due to fuel starvation. In the subsequent crash-landing one of the passengers was killed.

Flight details

The flight originated at Rand Airport and was destined first for Kimberley and then for Welkom in the Free State. On board were the pilot and two passengers who were all employees of the same company.

The aircraft left Rand Airport with a maximum load of fuel. The flight to Kimberley took 1 hour and 45 minutes. The pilot did not refuel at Kimberley as he had more than enough fuel for the onward journey to Welkom.

The onward journey included a quick flight over a housing development west of Kimberley, which added 18 kilometres to the trip. Flying time was calculated to be one hour.

As the aircraft approached Welkom Airport, the engine surged and then stopped. The pilot, who had been flying on the left-hand tank, now selected the right-hand tank and the fuel booster pump. The fuel gauge for the left-hand tank indicated that it was almost empty and that the right hand was a quarter full. He informed Welkom air traffic control (ATC) that his engine had failed but did not declare an emergency. No matter what the pilot attempted the engine could not be re-started.

At this stage the pilot positioned the aircraft in a glide, intending to carry out an emergency landing on a road. He was then in a left-hand turn and at about 500 feet above ground level. Nobody knows exactly what happened after that. Both the pilot and the surviving passenger received serious head injuries in the crash and could not recall the accident in full.

Investigators established that the aircraft had stalled while still quite high off the ground. Ground marks showed a low forward speed and a high vertical speed.

It is also possible that the seat belts were not tightened correctly. The rear passenger was thrown forward and out of the seat, resulting in fatal injury. The two front seats did not have upper-torso belts, so the occupants were thrown forward and received severe head injuries.

Significantly, it was discovered that there was enough fuel left in the right-hand tank for the pilot to land safely, had the engine not been starved.

Fuel exhaustion

Date:	18 October 2000, 06h15
Conditions:	Low-level cloud
Aircraft Type:	Cessna 182
Pilot Age:	41
Flying Hours:	289
Location:	Leydsdorp, Limpopo Province

Synopsis
On a flight between Komatipoort and Tzaneen the pilot realised that he did not have sufficient fuel and decided to make a precautionary landing. The Cessna overturned when its left wing hit a tree.

Flight details
The pilot took off from Komatipoort with two passengers on board. The pilot had not refuelled after his previous flight, calculating that he had enough fuel to get to Tzaneen with a 45 minute reserve.

The pilot was flying at 4 500 feet, but because of low cloud near Hoedspruit he asked air traffic control for clearance to fly low level from Hoedspruit to Tzaneen. Approximately 18 kilometres from Tzaneen, realising he was dangerously low on fuel, he decided to make a precautionary landing. While the pilot was looking for a suitable field in which to land, his aircraft ran out of fuel. The only landing option that remained was a dirt road. During the approach the wheels of the Cessna struck a telephone line and the plane's left wing hit a tree, causing the aircraft to overturn.

The pilot and his passengers were fortunate to escape with minor injuries. The pilot's fuel miscalculation could easily have had dire consequences.

The Cessna overturned when its left wing caught a tree during a precautionary landing.

No fuel, big fan stops!

Date:	18 June 2000, 12h30
Conditions:	Fine
Aircraft Type:	Robinson R-22
Pilot Age:	40
Flying Hours:	1017
Location:	3,6 km from Grand Central, Midrand, Gauteng

Synopsis

On a flight from Marble Hall in Mpumalanga to Grand Central Airport in Midrand, the R-22's engine died suddenly. The helicopter autorotated and crash-landed.

Flight details

The flight originated at Marble Hall and was destined for Indaba Heliport, just outside Johannesburg. The pilot was due to deliver the chopper to Grand Central Airport the following morning for maintenance. When he was close to his destination, the pilot called Indaba Heliport, but got no reply. He then decided to call Grand Central and ask for their joining and landing procedures.

The pilot stated that he had the airport in sight, when suddenly the chopper's engine died. He immediately selected a landing site and proceeded with an autorotation. As the pilot turned to avoid a clump of trees, the chopper's rotor speed decreased to a point where no lift was being generated. The chopper's descent was too fast and it plummeted into the ground. The main rotor severed the tail boom and the skids were badly contorted, as was the cabin area. The pilot sustained severe back injuries in the crash landing.

So what caused the accident?

The investigators found that the right-hand fuel tank was empty and the left tank contained 'an unusable amount of fuel'. On removing the carburettor, it became clear that the engine had been starved of fuel.

The Robinson R-22's tail boom was severed and its skids badly damaged.

CHAPTER 9

Less common factors

UP UNTIL this point, this book has dealt with fairly common causes of aircraft crashes.

Every so often, however, there are cases where less common factors result in aircraft tragedies. This chapter deals with some of these more unusual factors, including poor health, hypoxia, inadequate maintenance and possible sabotage.

Poor health

Date:	31 August 1991
Aircraft Type:	DH 82A
Pilot Age:	53
Flying Hours:	394
Location:	New Tempe Airport, Bloemfontein, Free State

Synopsis

On a short recreational flight the DH suffered engine failure, which sent it plunging to the ground before bursting into flames. Both the pilot and his passenger were killed.

Flight details

In trying to piece together the details of this tragic accident, let us briefly look at the pilot's history. Age 53, he held a valid pilot's licence and was type-rated. His logbook was filled in up until 18 February 1991, by which date he had accumulated 394 flying hours.

> When a pilot is assessed on his ability to fly a particular type of aircraft, he is said to be **type-rated.** He can only be assessed by another pilot who is type-rated on that particular aircraft.

Apparently he had flown regularly between April 1970 and January 1974. He then did not fly again until December 1990. Twice, between 1973 and 1975, he was required to submit a cardio-vascular report from a specialist. In 1977, it was decided that he should undergo an annual ECG. He did not have another medical examination until August 1982, when the ECG requirement was reinstated.

Several more ECG examinations were required between 1985 and 1990. The test results were normal. Consequently, the annual ECG requirement was altered to a biannual one.

The doctor who examined the pilot at his last ECG stated that he could find no cause for concern. He did however advise the pilot to stop his heavy smoking.

On the day of the crash, the pilot had loaded up a passenger for a short recreational flight. The aircraft took off on runway 10, made a right-hand turn and headed towards the hangar area.

A witness stated that as the aircraft passed a point between the north-south taxiway and the hangar car park area, it suddenly pitched up and levelled off at a height of 200 feet above ground level. The engine spluttered, then stopped completely. At this stage the aircraft pitched down into an excessive dive and plunged to the ground, narrowly missing a hangar. It burst into flames shortly after hitting the ground. The pilot and passenger were killed instantly.

Although the aircraft was virtually destroyed it was possible to ascertain that the flight controls were still operable. All cables and bracing wires were found attached.

An autopsy carried out on the pilot revealed that he had possibly suffered a minor thrombosis. This would have caused a period of intense pain, inducing involuntary muscular spasms throughout his whole body. This may have caused him to pull back on the throttle and push forward on the control column. It would also explain the spluttering of the engine and the slight pitching up of the aircraft before it plunged violently to the ground.

Hypoxia

Date:	12 December 1998, 13h45
Conditions:	Severe thunderstorms
Aircraft Type:	Piper PA-32R-301T
Pilot Age:	36
Flying Hours:	399.25
Location:	Rorke's Drift area, KwaZulu-Natal

Synopsis
The Piper was on a flight from Zeerust to Virginia Airport in Durban, when it encountered a severe storm in the Rorke's Drift area. Witnesses saw the aircraft breaking up in mid-air and falling to the ground.

Flight history
Before we look at the details of the flight, let's look at the condition of the aircraft itself. Certain components had been added to the airframe to improve airflow, but the modifications had not been properly certified by a maintenance organisation. The components had been well fitted, but it is possible that the modifications contributed to the destruction of the aircraft as will be seen later.

On the day of the flight, the pilot of the ill-fated plane phoned Durban weather office at approximately 10h00 to get a weather update. The flight plan called for a level of 13 500 feet, which was the minimum safety altitude (MSA) for an unpressurised aircraft.

The following is the transcript of the communication between the pilot of the Piper (LGI) and Durban Information (Control):

```
13:08    LGI:     Durban Information, LGI.
       Control:   LGI, Durban go ahead.
         LGI:     LGI from Zeerust to Virginia per
                  flight plan. Presently entering
                  your FIR-boundary at   Warden.
                  Maintaining level 135.LGI.
       Control:   Thank you LGI.(Unreadable word)
                  to Virginia, remaining strictly
                  VFR in sight of ground. QNH,
                  correction Squawk code 3067.
                  Report abeam Ladysmith.
         LGI:     Squawk code 3067, remain VFR and
                  report abeam Ladysmith.
       Control:   LGI.
13:09  Control:   LGI, Durban.
         LGI:     LGI.
       Control:   LGI, there is a line, there is a
                  lot of low cloud approximately
                  10 miles in from the coastline.
                  The cloud is on  the high ground.
                  I suggest at this stage you
                  route via the Tugela River, ehhm
                  position yourself to the, via
                  the Tugela River for Virginia.
         LGI:     Ehh, thank you for that advice
                  sir, we will route via the
                  Tugela River and position
                  ourselves on the Tugela River
                  for Virginia. LGI.
       Control:   LGI, just be advised of this
```

		(unreadable word) thunderstorms in the Estcourt/Mooi River area and could you look out for those as well.
	LGI:	Thank you sir. We will keep a look-out for the, uh thunderstorms en route.
13:21	**LGI:**	Durban, LGI.
	Control:	LGI, go ahead.
	LGI:	LGI, abeam LYV and request descent to level 115.
	Control:	LGI, descend flight level 115 and report maintaining.
	LGI:	Descend level 115 and report maintaining. LGI.
	Control:	LGI.
3:29	**LGI:**	Durban, LGI. Level 115, request descent to level, ehh, 95. LGI.
	Control:	LGI, Durban control (Unreadable word) flight level 95.
	LGI:	Ehh, say again is the descent approved. LGI.
	Control:	LGI, affirm, during traffic flight level 95. Report maintaining.
	LGI:	Descend 95 report maintaining, LGI.

The pilot did not respond to any of the calls from Durban Control after this point.

At an altitude of 13 500 feet, it is possible for occupants of aircraft to become hypoxic.

Hypoxia is a condition caused by lack of oxygen. It manifests in dizziness, headaches and extreme fatigue. Other symptoms include impaired vision and intellectual capacity, making it impossible for sufferers to comprehend their own disability. It thus results in poor judgement and delayed reaction time.

It is highly likely that the pilot, having flown for an hour and a half at 13 500 feet, was in fact hypoxic and unable to recognise the dangers in which he was about to place himself and his family.

By the time the aircraft got to Rorke's Drift a thunderstorm was raging. Witnesses on the ground reported seeing the aircraft entering a storm cloud. When it re-emerged, there were pieces falling from the aircraft after which it tumbled uncontrollably to the ground. The pilot and his family were killed.

Investigators found that the first part of the aircraft that broke off was the elevator trim tab which caused the stabilator to flutter and eventually fail. This resulted in an instant loss of longitudinal control causing the aircraft to pitch forward and then break up rapidly. This was not a survivable accident. It is probable that the aircraft would have broken up anyway because of the severe wind shear associated with the thunderstorm. The speed at which the pilot appeared to have lost control and the extreme altitude at which he had been flying for a long time, strongly suggest that he may have suffered from hypoxia.

Local villagers examine the wreckage of the Piper which lies on a hillside near Rorke's Drift.

A homebuilt Mustang

Date:	12 November 1999
Conditions:	Overcast, with a cloud base of 1 300 feet
Aircraft Type:	Mustang P51D
Pilot Age:	47
Flying Hours:	487
Location:	18 km east of George Airport, Western Cape

Synopsis

On a flight from Port Elizabeth to Bredasdorp, bad weather forced the pilot to divert to George. On landing, the wheels would not release and the pilot requested to route to the east of the field to try to lower the wheels manually. While attempting this he lost control and crashed.

Flight details

The Mustang, a two-thirds scale model of the real aircraft, was being flown from Port Elizabeth to Bredasdorp to participate in an air show the next day. As there was bad weather at Riversdale, the pilot elected to divert to George. He reported to George air traffic control (ATC) when he was abeam Mossel Bay. He was told to join right-hand downwind for runway 29. He reported downwind and was cleared for finals. Shortly afterwards, he called ATC and told them he had problems lowering the wheels. He requested to route east of the airport to try to fix the problem. Permission was given, which the pilot acknowledged. No further calls were heard and the wreckage was soon found east of the airport.

The following is a transcript of the contact between ATC and the Mustang:

```
06:19  Aircraft:   George - Uniform Lima Lima (ULL)
          Tower:   ULL  - George.
       Aircraft:   OK George, cloud is here
                   unbroken to the coastline. I am
                   now 26 miles from Riversdale.
```

I am going to descend low-level and suggest the aircraft fly low-level along the coast to Underberg – ULL.

Tower: Thank you. I have heard that and descend as required and if you could try to keep me updated reference so I can pass on to the other aircraft. Once you get to the coastline, let me know what it is like.

Aircraft: I can get contact with the aircraft – ULL.

06:26 **Tower:** ULL – George your position and altitude now.

06:28 **Tower:** ULL – George.

Aircraft: George calling ULL.

Tower: Affirm your position now and weather conditions there.
(No immediate response)
ULL if you read this transmission, contact Underberg 119.8 – good day.

06:30 **Aircraft:** George – ULL

Tower: ULL – George.

Aircraft: OK I am heading back to George. Low level the fog is getting worse at Riversdale so I am heading back low-level to George – ULL.

Tower: Thank you ULL. Continue down to George QNH 1019.

Aircraft: Copy 1019 and continue to George – ULL.

Tower: Are you at the coast at the moment?

Aircraft:		Negative, I am about 10 miles inland – ULL.
Tower:		Thank you ULL. Keep me advised.
Aircraft:		(Two clicks on the mike.)
06:36	**Aircraft:**	George –ULL. 27 miles to run – ULL.
Tower:		And your flight conditions at this time?
Aircraft:		The ceiling is descending all the time.
Tower:		OK. Keep me advised. Continue inbound. There is cloud south-west of the airfield at George, ahead of extended base. Sir, I have to estimate for your information that it will be at about 1000 feet. The north of the airfield towards the mountain is clear still but the cloud is moving in.
Aircraft:		OK. Copy that. Thank you very much.
06:42	**Aircraft:**	Tower ULL is abeam Mossel Bay with 12 miles to run.
Tower:		ULL join and report right-hand downwind for RWY 29. There is traffic in the circuit. Keep a look for the VFR traffic in the circuit.
Aircraft:		OK. Copy the traffic and will join right-hand downwind for 29.

Note: Circuit traffic advise the tower that the cloud base was 1 300 feet.

06:47	**Aircraft:**	George ULL. Coming up right-hand downwind and I have the

	traffic in sight, close to the field. I am about 3 miles out.
Tower:	Thank you ULL and report when ready for the base-leg turn. There is another aircraft. He is on base at the moment
Aircraft:	OK. Copy that I will call ready – ULL.
Tower:	Tango Victor standby break break – ULL the traffic in your inside is a Piper Cherokee, if you can follow that one. Position yourself behind the Piper Cherokee and report final RWY 29.
Aircraft:	I will keep a lookout for the traffic and will position behind the Cherokee – ULL.
06:49 **Aircraft:**	Tower – ULL.
Tower:	ULL – Tower.
Aircraft:	I would like to proceed to the east. I have a problem with the undercarriage I am going to release the undercarriage manually and I will call you when ready. I will maintain my position to the east over the coastline – ULL.
Tower:	I have no traffic out to the east. You advise when ready to return to the airfield and we will probably request a fly-past so we can check from the tower as well so maybe can help you once it's down – ULL.

This was the last communication between the Mustang and the George ATC tower, despite the tower trying three times to re-establish contact with the aircraft.

A witness saw the Mustang flying at an altitude of 700 to 800 feet. It suddenly pitched upwards as though the pilot was attempting an aerobatic manoeuvre and shortly afterwards the aircraft hit the ground. It struck the ground nose-down, with not much forward speed which suggests that the aircraft stalled. The pilot was killed instantly.

So, what went wrong?

Perhaps the answer lies in the fact that the aircraft had a history of landing gear failure. During the initial test flights numerous problems were encountered with the landing gear. In March 1989 the aircraft was landed with the wheels up. Modifications were made and the defects appeared to have been rectified.

The pilot admitted he had forgotten to lower the gear. Repairs were done by the original constructor.

In July 1994 the left-hand main undercarriage failed to extend. On landing the aircraft veered off the runway causing the right-hand gear to collapse.

The owner had carried out a number of uncertified modifications to the aircraft, one of which could have played a big part in its demise. This was the introduction of three 'machine gun' ports to the leading edge of the wings. Could this have changed the nature of the airflow over the wings and led the aircraft to stall?

According to the constructor and test pilot, the Mustang is a very demanding aircraft to fly. If you are not quick to recover – in one and a half turns – you will be in for the wildest ride of your life.

The ill-fated model Mustang.

Canopy latch!

Date:	28 August 1999, 07h55
Conditions:	Fine
Aircraft Type:	Aero Vodochody L39C
Pilot Age:	36
Flying Hours:	8 082
Location:	Matsapha Airport, Swaziland

Synopsis
During a simulated fly-past at an air show, the pilot of the Vodochody dived down to fly along the crowd line, which resulted in the rear canopy flying off. He immediately prepared to land, but realising that the engine was not reacting, had to crash-land the jet short of the runway.

95

Flight details

The pilot of the Vodochody had arrived at the Swaziland air show at Matsapha Airport at 05h55. He had asked to do a simulated display flight before the air show started.

At 07h52 he and his co-pilot took off, climbed to about 4 000 feet above ground level (AGL) and lined up to do a fast run along the crowd line. He then dived down to 400 feet AGL and as he reached the start of the crowd line, noticed that the engine noise had suddenly become louder. Realising that the rear canopy had become detached, he decided to land the L39 immediately. He put the engine into idle, reduced speed and prepared to land. When the speed reached 266 kilometres per hour he eased the throttle forward but soon realised that the engine was not responding. He decided to attempt a forced landing just short of the runway in an open field.

Just before the aircraft hit the ground the flaps were fully extended in an attempt to reduce speed. The aircraft hit the ground and skidded for about 50 metres. Both pilots exited the jet and waited for the emergency crew to arrive.

The Aero Vodochody lies broken in a field.

So what went wrong?

The investigators found evidence that the closing latches did not engage properly and that the canopy was just half-latched and not locked down. The rear canopy held until the aircraft was in a dive at which time speed and low pressure over the canopy caused it to lift and break into pieces. Debris from the canopy entered the engine causing the turbine blades to break. This rendered the engine powerless.

Although the two pilots survived the accident they sustained severe back injuries during the forced landing.

The rear canopy of the Vodochody was still attached, but broken.

Inadequate maintenance

Date:	4 January 1995
Conditions:	Fine, temperature 23°C
Aircraft Type:	Cessna 1721
Pilot Age:	22
Flying Hours:	130
Location:	Rand Airport, Gauteng

Synopsis
A light aircraft took off from Rand Airport but failed to gain sufficient height and turned back to land. However, before it could get back to the airport it collided with power lines and crashed.

Flight details
This was to be a recreational flight from Rand Airport to Pilanesberg in Limpopo Province. The pilot had been cleared for runway 29 and had been told to maintain 6 000 feet above ground level (AGL) and runway heading until passing the 'Silver Ball' (a spherical water reservoir used as a landmark by Rand air traffic control).

Shortly after take-off, ATC noticed that the Cessna was not climbing and asked the pilot if he was experiencing difficulties. The pilot replied that he was returning to Rand Airport as he could not gain height. The ATC then advised him to position for a left-hand base leg for runway 35. The pilot then appeared to be joining the circuit for runway 11. The ATC cleared him for that runway. Shortly afterwards the aircraft dipped and crashed.

> When a pilot comes in to land at an airport he will be told to turn onto a **left-** or **right-hand base leg,** (which is perpendicular to the runway), before he is cleared to head for the runway itself.

The following is the transcript between the aircraft (EZO) and Rand Airport control tower (TWR). KHW and FTI refer to other aircraft that were in the circuit at that time:

EZO:	Rand morning EZO.
TWR:	EZO good morning, go ahead.
EZO:	EZO requests taxi for a flight to Pilanesberg. We are four on board and request runway 29 for take-off.
TWR:	EZO have you filed a flight plan?
EZO:	Affirm we filed it under JBO, but it is just that the registration has changed.
TWR:	OK copied. JBO has now changed to EZO, taxi holding point 29, QNH 1026, confirm number on board.
EZO:	EZO is ready.
TWR:	EZO confirm you would like to turn out right or maintain runway heading?
EZO:	I'd like to turn out right, EZO.
TWR:	EZO cleared to take-off, maintain runway heading till past the Silver Ball then turn out right.
EZO:	Cleared for take-off, maintain runway heading till past the Silver Ball then turn out right, EZO.
TWR:	EZO maintaining 6000 feet.
EZO:	6000 feet, EZO.
TWR:	KHW keep a lookout for traffic

	rolling runway 29, maintaining runway heading and maintaining 6 000 feet.
KHW:	Copied that, I'll look out for the traffic, KHW.
TWR:	Thank you KHW. Maintain 6 500 feet on the downwind.
KHW:	Six five zero zero feet on the downwind, KHW.
TWR:	EZO traffic on downwind will be maintaining 6 500 feet.
EZO:	[unreadable]
TWR:	Report zone outbound.
EZO:	Report zone outbound, EZO.
TWR:	Confirm ops normal?

This is where the ATC realise that things are going wrong.

EZO:	Say again.
TWR:	Confirm everything is OK?
EZO:	I'm having a bit of a battle to climb here. I think I'm going to turn back if that's OK, EZO.
TWR:	EZO, confirm you are returning to the field?
EZO:	Affirm, EZO.
TWR:	EZO, you can return to the field, report on the base runway 35.
EZO:	Report base runway 35, EZO.
TWR:	EZO, would you prefer runway 11?
EZO:	I'll get back to you.
TWR:	EZO, Rand, are you on frequency?
TWR:	EZO, Rand, are you on frequency?
TWR:	KHW, make this a full-stop landing

FTI:	Rand tower, FTI over.
TWR:	There is an aircraft that has crashed. He is now by the Silver Ball. He's crashed at the Silver Ball just in the slope.

Reading the transcript, it becomes clear that ATC did everything in its power to get the aircraft back safely to Rand Airport. Unfortunately, the engine faltered just above a power pylon, hitting the earthing cable and the top conductors. The power line was conducting 275 000 volts, and there was evidence of electrical arcing on the right-hand wing. The aircraft caught fire on impact and crashed inverted into a boundary wall between two houses. On inspection it was found that the flying controls were correctly connected and secure.

So what prevented the aircraft from climbing correctly?

CAA investigators revealed that the aircraft had a very poor maintenance record. On dismantling the engine, excessive carbon deposits were found on the piston crowns and in the cylinder heads. These showed high oil consumption. The rings on the pistons showed signs of considerable gas 'blow-by'. This would explain the engine's inability to achieve its maximum power. In fact, the operator had exceeded the mandatory periodic inspection (MPI).

> All aircraft are subjected to a **mandatory periodic inspection (MPI)** by an approved aircraft maintenance organisation. This needs to be done every 100 flight hours or every year, depending which occurs first.

Recordkeeping was also found to be extremely poor. The operator stated that the aircraft had flown very little during the first half of December and a substantial amount in the second half of the month. In fact, the maintenance booklet salvaged from the wreckage showed that the aircraft had flown almost every day in December and had completed 82 hours in that month. However, these factors alone did not cause the crash.

Another contributing factor can be explained by a passage from the book *Aerodynamics for Naval Aviators*. It reads as follows:

> When an aircraft is flown below the minimum drag speed, it enters the region of reverse command, where an increase in power does not result in acceleration because of the induced drag associated with the higher angle of attack. In this region, if the aircraft is in level flight at a certain power setting, a decrease in the angle of attack will result in an increase in induced power and the aircraft will climb. Conversely, an increase in the angle of attack will result in a power deficiency and the aircraft will descend. Operation in the region of reverse command does not infer that greater control difficulty and dangerous conditions will exist. However, proper flying technique and precise control of the aircraft are most necessary for flight in this region. This phenomenon is called 'getting behind the power curve'.

So what does all this mean?

In order to avoid this accident the pilot would have had to fly the aircraft very carefully indeed – not gaining height by pulling on the controls, but flying smoothly with precise control and a steadier climb out.

He was not to know that the engine had gone beyond the normal flight time and that the aircraft was in fact not strictly airworthy. Also, he had gained most of his experience in more powerful aircraft at the coast, where the air is denser. He did not have much experience flying on the Highveld, where 'hot and high' conditions are more commonplace. He possibly failed to recognise the signs which indicated his aircraft was running out of air speed.

Sabotage!

Date:	1 March 1988, 15h28
Aircraft Type:	Embraer-110P1
Pilot Age:	38
Flying Hours:	5 684
Location:	12,6 km south-west of Johannesburg International, Gauteng

Synopsis

An aircraft on a commercial flight from Phalaborwa in Mpumalanga to Johannesburg International Airport was approximately 13 kilometres from touchdown when it exploded, killing all 17 people on board.

Flight details

Comair flight CAW206 was scheduled to fly from Phalaborwa to Johannesburg International Airport with two pilots and 15 passengers on board. The approach to Johannesburg International was progressing smoothly and communications between the pilot and Johannesburg International approach control showed no sign of the impending catastrophe. The Embraer was flying at a cruise altitude of 12 000 feet and was cleared to 9 000 feet.

The pilot was told that he was second in line to approach the runway with a slower aircraft in front of him. He was given the choice of runway 33 or 03-left. If he remained lined up on 33, then he would need to slow down. As there was reduced visibility the pilot decided to choose 03-left so as not to worry about the slower aircraft. Approach control cleared him for 8 000 and then 7 500 feet. It was around this time that he was cleared for an instrument landing system (ILS) approach. The aircraft was seen on the approach radar to establish the ILS glide slope at 7 500 feet and was 13 kilometres out. The ATC said 'cheers' to the aircraft, but got no reply.

It was at this precise moment that the Embraer was blown out of the sky.

Several witnesses in the area reported hearing a huge explosion. What they saw will probably haunt them for the rest of their lives. Nearly every witness gave a similar account of what followed. They saw a cloud of smoke around the aircraft, and then the front section of fuselage breaking away and falling to earth. The body of the aircraft was then seen spiralling down in a slight nose-down attitude. Once it hit the ground it exploded into flames. There were no survivors.

The aircraft exploded over an industrial area but luckily no buildings were hit by fuselage wreckage. This fell onto the main gate of a nearby factory, damaging it badly. Falling debris slightly damaged other buildings in the vicinity.

Nobody could fathom what had caused the accident. Once investigators started their inspection of the wreckage, they soon became aware of significant irregularities.

For instance, extensive damage to two of the seats and injuries to the passengers showed that a violent explosion had occurred. The fact that the nose section had been severed also supported this. Another point to consider was that the debris had been scattered over a wide area – consistent with a break-up in the air. The forensic police were called in and tests were performed on various parts of the wreckage. The forensic team confirmed that the two badly damaged passenger seats showed high concentrations of nitroglycerine and ammonium nitrate. The combination of these two ingredients makes up a powerful bomb.

In 1988 security was non-existent at a minor airport like Phalaborwa. Passengers were only required to identify their luggage before boarding the aircraft and it was then loaded into the hold. No screening equipment was available and hand luggage was never inspected. Smuggling a bomb on to such a flight would have been a simple task. Police were never able to establish which of the passengers detonated the bomb that claimed 17 lives.

Runway blues

Date:	2 May 1999
Conditions:	Fine
Aircraft Type:	Piper 140
Pilot Age:	26
Flying Hours:	98
Location:	Cathedral Peak, Drakensberg, KwaZulu-Natal

Synopsis
Flying into a very tight runway, the pilot of the Piper applied brakes but this appeared to have no effect. He then applied the parking brake which caused the aircraft to veer to the right and hit an embankment.

Flight details

The flight originated fom Brakpan Airport in Gauteng. The Piper's pilot was relatively inexperienced. He was unfamiliar with the aircraft and it was his first flight into the Cathedral Peak area.

The runway at Cathedral Peak is very difficult to land on. You can land only one way and take off another. It is not, in my opinion, the kind of airstrip for anyone but the most experienced flyer.

However, to the pilot's credit, he did inspect the runway by flying overhead and confirming that it was clear of any obstacles. He selected full flaps and landed on the grass runway, immediately applying brakes as the end was approaching rapidly. Suspecting that the brakes were not working properly, he decided to apply the parking brake. The aircraft ran over an uneven stretch of ground and at this point the pilot lost control of the aircraft. It veered strongly to the right, its right-hand wing impacting an embankment.

Investigators found that the pilot had touched down very deeply on the runway and the wheels had skidded all the way from point of contact to point of impact! As with a car, when the wheels are turning one has maximum braking power. However, if the wheels are locked and skidding there is very little or no braking effect.

Investigators recommended to the owners that the runway be closed.

As can be seen in the photo below, the runway formed part of a golf course. Buildings in the run-off area posed a daunting safety hazard.

The Cathedral Peak runway. Note the proximity of the buildings to the run-off area.

The Piper lies marooned on a bank, overlooking the Cathedral Peak golf course.

Blow me down

Date:	20 April 2001, 06h47
Conditions:	Fine
Aircraft Type:	Cessna 150J
Pilot Age:	25
Flying Hours:	2 500
Location:	Cape Town International Airport, Western Cape

Synopsis
As a Cessna 150 taxied towards the runway, a Boeing 747 throttled up to move forward to the correct parking bay, blowing the Cessna over.

Details
Although not technically a flight, this unusual accident deserves a mention.

A British Airways Boeing 747 had just landed on runway 19 and was taxiing towards Apron A (see figure 1 below).

A Cessna 150 with an instructor and a student pilot on board was on taxiway H. Ground control told them to hold until the 747 passed them. After the 747 passed them and was on B1, the Cessna started to follow. The 747 proceeded onto Apron B then onto Apron A. The 747 then turned left into A6 (see figure 2). At this stage the Cessna stopped short and waited for the 747 to stop. As the 747 stopped, the Cessna moved forward. It was then that the 747 increased power to inch forward to the correct parking bay. The Cessna became caught in the considerable jet blast and overturned.

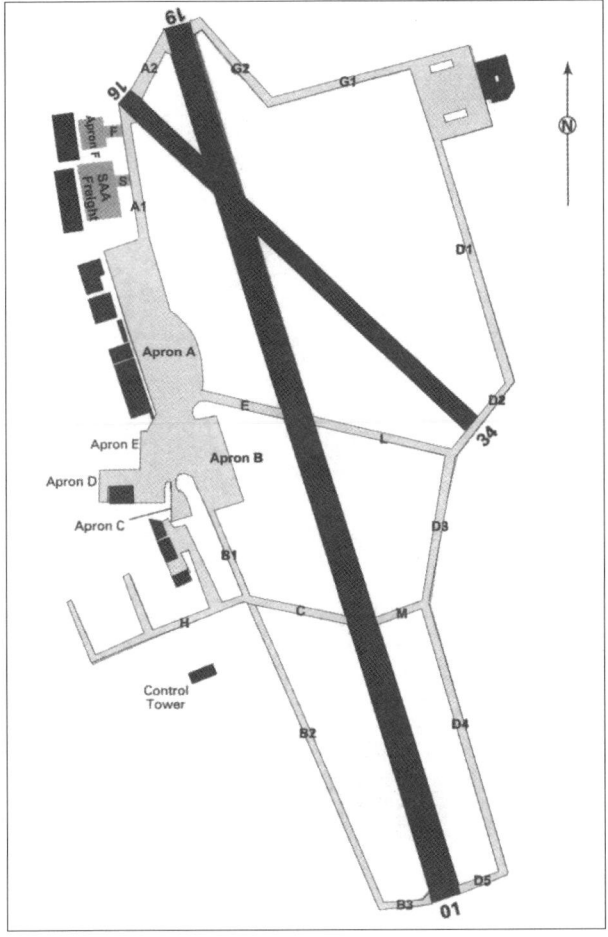

Figure 1

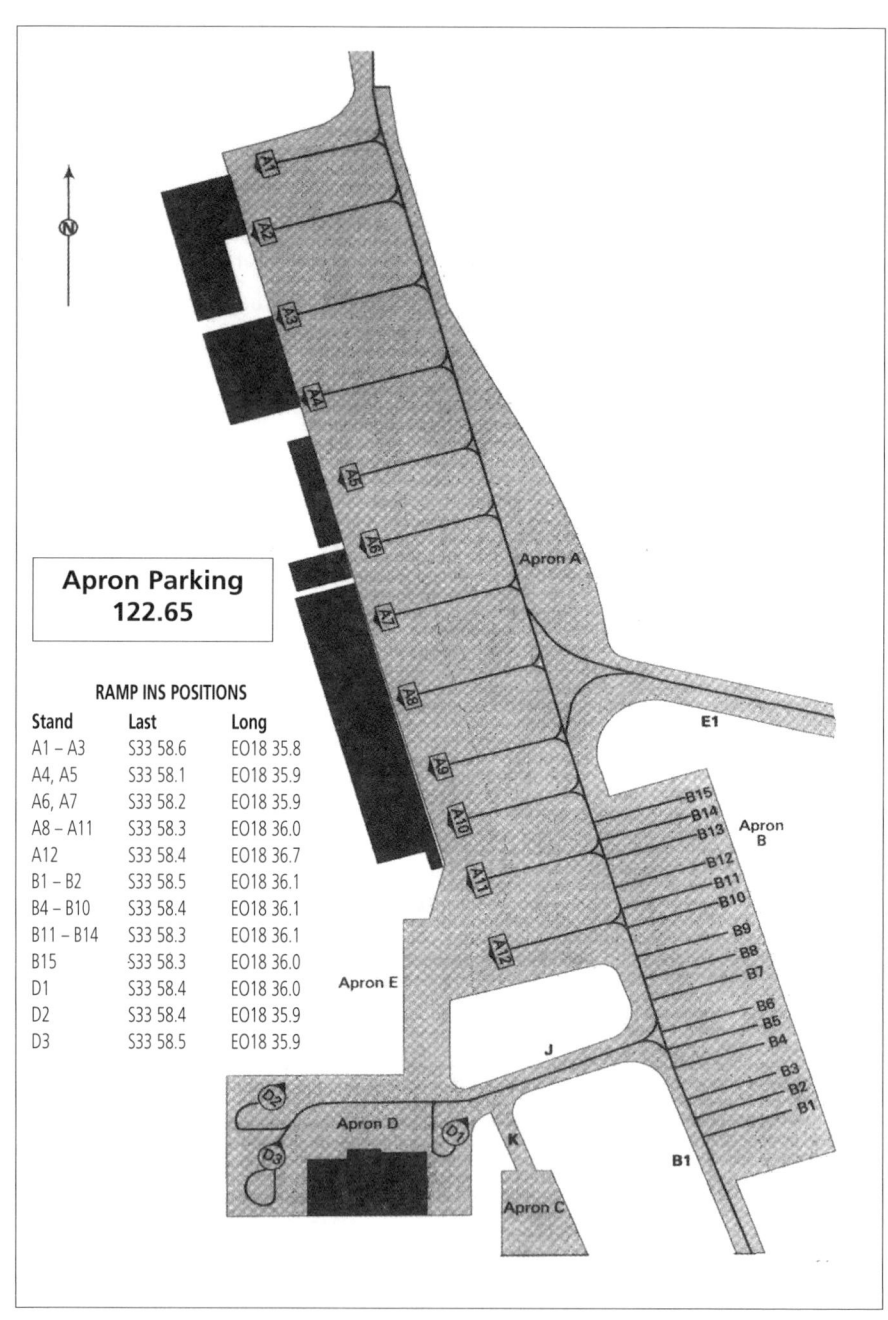

**Apron Parking
122.65**

RAMP INS POSITIONS

Stand	Last	Long
A1 – A3	S33 58.6	EO18 35.8
A4, A5	S33 58.1	EO18 35.9
A6, A7	S33 58.2	EO18 35.9
A8 – A11	S33 58.3	EO18 36.0
A12	S33 58.4	EO18 36.7
B1 – B2	S33 58.5	EO18 36.1
B4 – B10	S33 58.4	EO18 36.1
B11 – B14	S33 58.3	EO18 36.1
B15	S33 58.3	EO18 36.0
D1	S33 58.4	EO18 36.0
D2	S33 58.4	EO18 35.9
D3	S33 58.5	EO18 35.9

Figure 2

The following is the transcript between ground traffic control, the 747 (GC-IVE), the Cessna (UYS) and a Cessna 172 (KSV) who was taxiing to the holding point of runway 19:

06:37:27   **UYS:**   Cape Town Ground, UYS good morning.

06:37:30 **Ground:**   UYS Ground, a very good morning.

06:37:32   **UYS:**   UYS, eeehh 150 with 2 people on board for a flight to the Delta 200.

06:37:39 **Ground:**   UYS the QNH 1017, taxi to the holding point on Hotel.

06:37:44   **UYS:**   QNH 1017, taxi to the holding point on Hotel, UYS.

06:37:50 **GC-IVE:**   Ground good morning, Speedbird 59 clear runway 19.

06:37:55 **Ground:**   Speedbird 59 a very good day to you. Welcome to Cape Town. Continue taxi on Bravo to the ramp, your gate and registration please.

06:38:00 **GC-IVE:**   A-6 on Bravo to the ramp GC-IVE, lovely view.

06:38:08 **Ground:**   Speedbird 59, glad you enjoyed it, enjoy your stay.

06:38:11 **GC-IVE:**   Thank you.

06:39:28   **UYS:**   Cape Town Ground, UYS, eh, please confirm that we can taxi to holding point on Hotel.

06:39:33 **Ground:**   Affirm UYS.

06:39:36   **UYS:**   Thank you.

06:39:40 **Ground:**   UYS you will be, there will be a 747 passing on Bravo, you are to follow it via Bravo Alfa holding point runway 19.

06:39:48   **UYS:**   We'll follow the traffic via

		Bravo Alfa to holding point runway 19, UYS.
06:41:08	KSV:	Good morning Cape Town Ground, KSV.
06:41:10	Ground:	KSV, Ground, a very good morning.
06:41:32	KSV:	KSV is a Charlie 172 and a training flight from Cape Aero Club to Ysterfontein returning, two on board, in-flight VX received, QNH 1017, KSV.
06:41:48	Ground:	KSV, QNH is 1017, taxi Hotel Bravo Alfa holding point runway 19.
06:41:56	KSV:	Taxi holding point, eehh, Hotel Bravo Alfa, holding point 19, KSV.
06:43:25	Ground:	UYS at the holding point when ready, contact tower frequency 118,1, good day.
06:44:11	UYS:	We'll contact tower on 118,1 when ready, KSV .
06:44:48	Ground:	KSV, if possible could you stop your aircraft right there and without endangering yourself, see if you could help the occupants of the light aircraft, the fire department is on its way.
06:44:56	KSV:	KSV.
06:45:20	Ground:	T2-expedite your vehicles to the ramp there is a Cessna 150 with two occupants that has flipped behind one of the aircraft, there is also a person walking out to try and help them.

The Cessna came to rest 70 metres behind the 747. Fortunately, neither of the occupants were injured in the accident.

The Cessna was directly in line with the jet blast of the Boeing 747.

CHAPTER 10

The CAA

THE Civil Aviation Authority (CAA) was established on 1 October 1988. It is a stand alone authority, charged with promoting, regulating and enforcing civil aviation safety and security.

Its areas of responsibility are widespread: they include certification of airports and heliports, and overseeing the examining, licensing and training of aviation personnel.

A large part of their work includes examining aircraft crash sites. This is not a job for the faint-hearted. They are often asked to go scrambling up a mountainside to retrieve bits and pieces of wreckage. Sometimes they are faced with the grisly task of examining wreckage where the bodies of the victims are still trapped inside.

The CAA issues a monthly publication called *The Safety Link,* which discusses a range of interesting aviation topics. With their permission, I'd like to reproduce one such article. It provides a fascinating insight into the crucial, sometime split-second decisions that pilots have to make in emergency situations.

Pilots – It's your decision

Adverse weather: Probably the most important decision the pilot in general aviation will face at some time or another is whether he should continue to fly in adverse weather or not. Over 80% of CFIT accidents are due to bad weather conditions.

Weather conditions can change very rapidly, so the weather forecast could help to start with, but by the time a pilot reaches his destination, things could have changed drastically. Long before the pilot gets himself into trouble, the question should be, 'Should I carry on, or do I divert?' If the weather is bad and you are not instrument-rated, then there should be no option – turn back while you can! It does not reflect badly on your abilities as a pilot if you turn back. In fact, it shows good judgement. It doesn't matter what the circumstances are, urgent meeting, got to get back to work on Monday etc. If you are not happy, turn around.

Plan ahead, carry your driver's license and credit card to enable you to hire a car and drive or book into a hotel in a different town. It will reflect badly if you bend the aircraft or even worse, kill yourself!

But I've done it before: is no guarantee that it will work again! Perhaps you flew into bad weather last time and flew by instinct and landed safely. Think again! You were probably incredibly lucky last time. Look at it another way. If someone close to you said to you they were going to blindfold themselves then walk across the highway and hope not to get knocked down – just because they tried yesterday and succeeded – what would you say to them?

But I know someone else who did it: The fact that someone else can do it does not mean that you can – he or she might have better skills or more training than you. Assess your limitations – then live by them. Do not compare yourself to other pilots.

But you promised: Never promise someone a flight for a specific day or time. You could be pushed into a situation in which your sound judgement is taken out of your hands. This is especially true when it comes to a promise made to a child. You don't want to let them down or it's not easy to say no.

Peer pressure: Don't let other pilots or even passengers pressure you into changing your mind or decision. If you have paying passengers and you do not think the weather is good enough to fly, postpone the flight. Tell them it's better to get there one hour or even one day late than to not get there at all. If they then go to someone else who will fly them, so be it.

Audiences: More than half the fatal accidents involving aerobatics or low flying involved an audience. They seldom occur at air shows, but more at club airfields or just to impress friends on the ground. Most people on the ground wonder when you are going to crash and ask themselves why you have a licence at all.

Statistics confirm that it is more often than not young male pilots who are guilty of illegal aerobatics. Again, statistics tell us that the older pilot is the one more likely to fly into terrain (CFIT).

Pilots that are involved in aerobatic or low flying accidents are usually highly experienced. Perhaps they believe that their high flying hours preclude them from flying into the ground.

Pilots who are involved in controlled flight into terrain (CFIT) are also usually highly experienced. Do they also believe that their hours equip them with special powers?

Pilots with few hours (student pilots and just qualified) are less likely to be involved in an accident. Are they more aware and more careful? Pilots with hours between 200–500 are more likely to be involved with accidents. Is that because they are moving to more ambitious flying?

Speak to most pilots and they will impart a super confident feeling. 'Trust me, I'm a pilot', as if they are superhuman. The fact is that pilots are human and *do* make mistakes.

Check your instruments: Just as pilots can (and do) make mistakes, aircraft and instruments sometimes fail. It is important to check instruments regularly. Do not assume that your attention will be drawn to a problem. You need to check your instruments all the time. For example, your fuel gauge might be stuck on full. You think 'all OK – plenty of fuel', until the propeller stops, then you panic! Check for movement of the needle. If it is working, it will move around.

A GPS is becoming a common accessory for pilots. Although it can be a very useful tool and is probably an overall safety plus, a few words of caution:

- Never use GPS as your primary means of navigation.
- Never use it to land in bad visibility.
- Do not spend time 'head down', not watching the the world outside.
- Do not have blind faith in GPS data. GPS's are not infallible and can be faulty.
- Never fly in conditions you would normally avoid, just because you have GPS. It will not reduce the risks!

CHAPTER 11

The experts

I APPROACHED THREE of the country's top pilots for their top five safety tips.

Karl Jensen (pictured above) is an experienced pilot with 24 511 flying hours behind him – many of those hours at the controls of a Boeing 747 as senior captain. Although recently retired from SAA, his career stretches back to 1962 when he started flying with the South African Air Force. He is a man whose passion for flying is second to none and I do not for one minute believe we have heard the last of Karl as a pilot! He has recently been made National Flying Safety Officer of the Experimental Aircraft Association, which is the biggest aviation organisation in SA for flying enthusiasts. His safety tips are as follows:

1. Take your training seriously and maintain humility to that training. Most flying training has evolved from a scientific background that has been developed by dedicated people who have safety as their

first priority. Keep abreast of the latest developments in flying training. Flying training in aviation is forever ongoing – remember you never know it all!

2. Spatial awareness. Always look out and anticipate what is going to happen next – it's no good being the first to arrive at an accident because you didn't anticipate the consequences of your actions or the trajectory of your aircraft and others about you. Anticipate that others around you are unaware of your presence. In all forms of aviation, an awareness of an alternative or forced landing field should always be borne in mind.

3. Be respectful of weather. A sound knowledge of meteorology is a prerequisite for a safe flight. Should the weather not conform to the forecast, turn back or land. 'Pressonitis' is a fatal disease. Meteorology is not an exact science and without disrespect to the profession of meteorologists, weather forecasts should be treated as a horoscope of numbers if the forecast is good and respected when the forecast is bad.

4. While a degree of confidence is a prerequisite to comfortably fly an aircraft, malaise, overconfidence, aggression, fanaticism, disrespect for property and other people, are certain recipes for disaster. Appreciate your limitations and be mindful of them.

5. Thorough preparation for a flight is essential. The pre-flight preparation includes adequate training, physical and mental preparedness, aircraft serviceability including adequate fuel and oil, observing aircraft performance limitations and a love of aviation!

Scully Levin is one of the country's finest aerobatic pilots, with a vast amount of flying time behind him. He is the owner and leader of the Shurlok Pitts Special team, and also leads the Nissan Harvard team.

And this is just his part-time, fun flying! On top of all this, Scully is a Boeing 747 captain, flying to numerous overseas destinations. In between all of this he is a family man who finds time to answer nagging questions from people like me! A man of vast experience, his safety tips to pilots are:

1. Know your own limitations.

2. Maintain situational awareness at all times.

3. Don't ever let an aircraft take you to some place that your brain has not arrived at a couple of minutes earlier.

4. Air speed is cash in hand. Height is money in the bank.

5. Nobody who gets too relaxed builds up much flying time.

Stu Davidson is another of South Africa's aviation personalities. He can be seen at air shows around the country, flying interesting aircraft. From the Boeing Stearsman Crunchie, wing-walking aircraft to a Hawker Sea Fury, Stu has flown in them all. He has a passion for flying and currently has in excess of 5 500 hours. Based in Port Elizabeth, Stu is a true sports pilot, who owns one of the best aerobatic aircraft in the world – the SU-29 Sukhoi. The following are Stu's safety tips:

1. Stay within your limits – and those of the aircraft.

2. Do not take short cuts with maintenance.

3. Always stay healthy and be in the right frame of mind.

4. Do not drink heavily the night before a flight.

5. Know the height gain and height loss in aerobatic manoeuvres, i.e. know your aircraft.

Dedication

The sea claims lives

Date:	30 December 1997
Conditions:	Fine
Aircraft Type:	PA28-180
Pilot Age:	24 and 18
Flying Hours:	914 and 75
Location:	Kenton-on-Sea, Eastern Cape

Synopsis
A training conversion flight ended in tragedy for a student pilot and her instructor, when their aircraft stalled and spun into the sea.

Flight details
The aircraft was on a conversion training exercise off Kenton-on-Sea, a popular holiday resort on the Eastern Cape coast. Witnesses stated that they saw the aircraft nosedive, spin and crash into the sea. The plane sank almost immediately and because of the very strong current off the Kenton coast, it was never found. This of course made it very difficult to diagnose the precise cause of the accident. The only clues came from a few witnesses who said that the aircraft stalled with insufficient height to recover, rather than breaking up mid-air.

Above the headland at Shelly Beach, Kenton-on-Sea, is a memorial stone dedicated to the student pilot, a young Italian girl named Marcella Marinelli. The photograph below was taken on 30 December 2001, exactly four years after Marcella died.

It was very moving to stand in this spot and view this memorial to a young person who had a passion for flying and died doing what she enjoyed most.

Even though I didn't know her, I was struck by the pain that her family must have felt on learning of her death.

I'd therefore like to dedicate this book not only to Marcella, but to all the other pilots and passengers mentioned in this book, who lost their lives so tragically.

May they rest in peace.

Marcella Marinella's memorial stone lies on a headland at Shelley Beach, Kenton-on-Sea.

Glossary

ADF	Automatic direction finder
AGL	Above ground level. Heights of clouds reported in AGL
Aerad guide	A book with every airport/aerodrome in it with landing details
ALERFA	The second alert for suspected missing aircraft
AMSL	Above mean sea level
ASL	Above sea level
ATC	Air traffic controller
CATCO	Chief air traffic control/controller
CFIT	Controlled flight into terrain
DETRESFA	The third alert phase
DME	Distance measuring equipment
ETA	Estimated time of arrival
FIR	Flight information region
GPS	Global positioning system
IFR	Instrument flight rating
ILS	Instrument landing system
IMC	Instrument meteorological conditions
INCERFA	First alert of suspected missing aircraft – 30 minutes after ETA
Jeppeson guide	An airport flight guide explaining runway and landing details
Knots	Nautical speed. 100 knots equals 185 kph
MPI	Mandatory periodic inspection
NDB	Non-directional radio beacon – ADF points to this beacon
PPL	Private pilot's licence
rpm	Revs per minute
QNH	Altimeter setting – set in Hg or millibars
STOL	Short take-off and landing aircraft
VFR	Visual flight rules
VMC	Visual meteorological conditions
VOR	Very high frequency omnidirectional radio

*This book is dedicated to Aggie – the best mother
a guy could have.*